Urban Sustainability

The Urban Sustainability Book Series is a valuable resource for sustainability and urban-related education and research. It offers an inter-disciplinary platform covering all four areas of practice, policy, education, research, and their nexus. The publications in this series are related to critical areas of sustainability, urban studies, planning, and urban geography.

This book series aims to put together cutting-edge research findings linked to the overarching field of urban sustainability. The scope and nature of the topic are broad and interdisciplinary and bring together various associated disciplines from sustainable development, environmental sciences, urbanism, etc. With many advanced research findings in the field, there is a need to put together various discussions and contributions on specific sustainability fields, covering a good range of topics on sustainable development, sustainable urbanism, and urban sustainability. Despite the broad range of issues, we note the importance of practical and policy-oriented directions, extending the literature and directions and pathways towards achieving urban sustainability.

The series will appeal to urbanists, geographers, planners, engineers, architects, governmental authorities, policymakers, researchers of all levels, and to all of those interested in a wide-ranging overview of urban sustainability and its associated fields. The series includes monographs and edited volumes, covering a range of topics under the urban sustainability topic, which can also be used for teaching materials.

Yongping Zhang

Understanding Shared Micromobility in China Using Massive User-Generated Trip Data

Springer

Yongping Zhang
School of Public Affairs
Zhejiang University
Hangzhou, Zhejiang, China

ISSN 2731-6483 ISSN 2731-6491 (electronic)
Urban Sustainability
ISBN 978-981-96-8491-5 ISBN 978-981-96-8492-2 (eBook)
https://doi.org/10.1007/978-981-96-8492-2

This Springer imprint is published by the registered company Springer Nature Singapore Pte Ltd.
The registered company address is: 152 Beach Road, #21-01/04 Gateway East, Singapore 189721, Singapore

Preface

Shared micromobility, such as dockless bike-sharing and e-bike-sharing, has rapidly expanded since 2015 in China. Unlike traditional public bike systems, the emerging dockless services eliminate the need for fixed docking stations, allowing users to locate, unlock, and pay for shared bikes/e-bikes via mobile apps, and then park them near their destinations. Over the past decade, shared micromobility has been increasingly integrated into urban transport networks in most Chinese cities, from the Tier-1 cities (e.g., Beijing and Shanghai) to many small and medium-sized cities (e.g., Jinhua and Xiangshan) that may not be very familiar by international readers. It offers flexible, low-emission options for short-distance travel, enhancing multi-modal connectivity, and also contributes to congestion mitigation, public health promotion, and the creation of citizen-friendly travel environments. *The Outline for Building a Strong Transport Nation*, which was issued in September 2019 and by the Central Committee of the Communist Party of China and the State Council, perhaps is the most important guidance to develop China's transport system. It emphasised the need to "vigorously develop shared transport and build a service system based on mobile smart terminal technology" and "improve urban pedestrian and non-motorised transport systems and improve the quality of walking, cycling and other travel." This (partly) demonstrates the important role of shared micromobility in the future transport system of China.

The development of shared micromobility has created new opportunities for data-driven transport research by providing large volumes of passively collected and user-generated trip data. Unlike traditional data sources such as household travel surveys and census statistics, these datasets are typically collected automatically through GPS-enabled devices embedded in shared bikes or e-bikes and are timestamped at fine temporal intervals. As such, they provide real-time or near real-time records of millions of individual trips, often including origin-destination pairs, route traces, trip durations, and speeds. Massive user-generated trip data of shared micromobility allows researchers to observe collective behaviours and mobility dynamics across entire urban systems. The spatiotemporal granularity of the data facilitates the analysis of fine-scale phenomena, such as the last-mile connectivity and the diurnal rhythms of travel demand. Moreover, the data contributes to a better understanding of the interplays between shared micromobility and other urban elements (e.g., land uses and residents).

Against this background, this book aims to understand shared micromobility in China using massive user-generated trip data. The trip datasets used in this book are mainly provided by Hellobike, which is China's leading service provider. According to Hellobike's *2023 Sustainable Development and ESG Report*, released on June 5, 2024, it had over 600 million registered users by the end of 2023. Its shared micromobility service has expanded to over 500 Chinese cities. The cumulative riding distance reached 42 billion kilometres for shared bikes and 16 billion kilometres for shared e-bikes. This demonstrates the representativeness of the service provider and the usefulness of its data.

Particularly, there are seven chapters in this book. Chapters 1, 2 and 3 focus on understanding travel characteristics of shared micromobility. Specifically, Chapter 1 identifies biking islands in Shanghai based on the theory of percolation. Here, biking islands is a new concept proposed to describe distinct geographical areas attracting cyclists and characterised by a high density of cycling activities. Chapter 2 examines cross-city travel behaviour and its associated cycling spatial structures in Zhejiang Province using a network-based approach. Chapter 3 infers trip purposes of shared micromobility in Ningbo City using an analytical framework featured by the Latent Dirichlet Allocation model. Chapters 4, 5, and 6 discuss the external impacts of shared micromobility by considering its environmental and equity impacts and its relationship with 15-minute cities. For Chapter 4, it quantifies the potential of carbon emission reductions from shared bikes and e-bikes in three Chinese cities at varying stages of economic development. Chapter 5 aims to evaluate the equity of dockless bike-sharing services in Wenzhou City. It relies on an introduction of an enhanced accessibility model, termed service-opportunity accessibility, which considers both the attractiveness and accessibility of urban opportunities for shared micromobility users. Chapter 6 investigates the potential contribution of shared bikes to the development of 15-minute cities in the whole province of Zhejiang to offer valuable insights for transport planning and policymaking. Finally, Chapter 7 proposes a series of methods to support electric fence planning for dockless bike-sharing services.

The series can be used as a reference for geographers, urban planners, transport professionals and policymakers. It also has the potential to be used as supplementary teaching materials for undergraduate and graduate students in transport, geography, urban planning and policy.

The publication of this book is financially supported by the Zhejiang Provincial Natural Science Foundation of China (LY24D010001), the Humanities and Social Science Fund of the Ministry of Education of China (23YJC630244) and Natural Science Foundation of China (72574195). I must thank my collaborators and graduate students for their kind support and assistance. I must thank the publishers (e.g., Elsevier) who authorised me to include the published peer-reviewed journal papers in this book. I also would like to express my gratitude to the editors of Springer for their understanding and support, making it possible to publish this book.

Hangzhou, China
September 2025

Yongping Zhang

Contents

About the Author

Yongping Zhang is a tenure-track assistant professor at the School of Public Affairs, Zhejiang University, China. He also serves as the deputy director of the ZJU-CMZJ Joint Lab on Data Intelligence and Urban Future, the director of the City+ Research Network (LinkedIn: City+; Wechat account: CityPlus2016; RedNotebook: 5049418472), and the editorial board member of the *Journal of Transport Geography*. He has a research interest in urban big data analysis, smart cities, digital urban governance, sustainable transport and GIS, with a focus on understanding various urban issues by emerging quantitative methods and urban data sources. At present, he has published more than 40 peer-reviewed papers in international journals such as *Applied Energy*, *Cities*, *Environment and Planning B*, *Journal of Transport Geography*, *Sustainable Cities and Society, and Transportation Research Part D*. He received his PhD degree in Advanced Spatial Analysis from University College London (UCL), the United Kingdom. Before this, he obtained three master degrees: one in European Spatial Planning and Environmental Policy from Cardiff University (the UK), one in Environment and Society Studies from Radboud University Nijmegen (the Netherlands), and one in Urban and Regional Planning from Peking University (China). He obtained his bachelor degree in Geographic Information System (GIS) from Wuhan University (China).

Chapter 1
Identify Biking Islands Using the Theory of Percolation and Cycling Trajectory Data

1.1 Introduction

As a critical component of the multimodal transport ecosystem, cycling generates a lot of positive outcomes for urban development, like mitigating traffic congestion and reducing energy consumption as well as the risk of chronic diseases (e.g., obesity) [8, 29]. In conjunction with the growing popularity of shared economy, bike-sharing, which provides users with the flexibility to make short-term bike rentals, is playing a more and more important role in promoting cycling mode in cities worldwide. Many studies have been conducted on bike-sharing systems in recent years, focusing on analysing the characteristics of the system (e.g., usage pattern and cycling behaviour classification) [18], developing mathematical models for rebalancing bike-sharing loads within a network [9, 19, 21], or optimising bike-sharing infrastructure planning [10, 20]. However, the majority of existing studies have limited ability to unveil spatial distribution (or structure) of real bike usage, reflected by their actual trajectories. For example, it is relatively easy to identify the bike stations that are frequently used but usually difficult to find out the bike lanes or geographical areas that are more attractive or suitable for cycling [20]. This limitation is mainly because most of the established bike-sharing services are dock-based, where users need to check out the bicycle at one station and return it at another designated station within the bike-sharing network. However, transient destinations and/or the intermediate routes between the origin and destination stations of a trip are unknown. Correspondingly, no spatiotemporal information about where and when people travel between the start and end stations is available. It is therefore unlikely to accurately delineate the fuller picture of the bike usage's spatial distribution.

Since dockless bikes are GPS-enabled, data generated by dockless bike-sharing services are bike trajectories, which accurately record spatiotemporal information of bike trips and thereby offer great opportunities to understand people's bike usage or cycling behaviour. Considering this, the objective of this chapter is to develop a new concept, i.e., biking islands, defined as geographical areas of interest with

Y. Zhang, *Understanding Shared Micromobility in China Using Massive User-Generated Trip Data*, Urban Sustainability,
https://doi.org/10.1007/978-981-96-8492-2_1

a high concentration of bike usage. Certain areas within a city may have mature bike facilities (e.g., bike lanes and parking spaces), making cycling in these areas more appealing and safer. As a result, compared to other areas, they may generate more bike usage or attract more people to ride a bike. Typical places include catchment areas of metro stations, university campuses and large-scale communities with well-developed bike facilities. The reason this chapter coins this new concept as biking islands is that areas with a high concentration of bike usage in a city resemble islands in the ocean. Identifying biking islands is helpful in understanding the travel behaviour of cyclists and urban structures used for cycling. The identification can also support relevant urban and transport planning. Non-motorised transport, including walking and bicycling, is often regarded as a key element of sustainable urban transport [3]. To encourage non-motorised transport, the identified biking islands have the potential to serve as designated areas for cycling where sufficient bike facilities are provided, and motorised transport modes are restricted or even prohibited to ensure the convenience and safety of cyclists and support the development of bike-friendly cities. On the other hand, the rapid development of dockless bike-sharing services not only generates considerable socioeconomic and environmental benefits but also causes a variety of new urban issues, such as illegal parking and dangerous cycling behaviour [27]. Identifying biking islands is helpful in finding out areas where dockless service should be encouraged or regulated, with joint consideration of the built environment, and urban plans and policies. An extra example is that it is also possible to identify the critical road segments that impede the biking flows between biking islands or of the entire transport network. Improving the biking facilities in these critical road segments is of great importance to improve the efficiency of the whole cycling system.

Because of the ability to accurately delineate the user's cycling behaviour, bike trajectory data are well suited to quantitatively measure the level of bike usage in cities. An intuitive way to detect areas with a high concentration of bike usage is to classify the raw bike trajectories into different clusters. However, the clusters obtained may overlap with each other in the geographical space, and the elements (i.e., trajectories or trajectory partitions) within one cluster may be geographically disconnected [13]. To overcome these issues, this chapter first aligns the raw trajectories onto the road network via map-matching algorithms [22] and then applies percolation theory to identify biking islands from the road network, which is weighted by bike usage. Percolation theory is selected because of its suitability in identifying various clusters (such as cities and local traffic flows) in an urban context [2, 14]. Specifically, the theory can be used to understand the behaviour of clusters in a graph, and can deal with the emergence of clusters of different sizes [11]. In addition, compared to other clustering algorithms, such as DBSCAN (density-based spatial clustering of applications with noise) [28] and community detection algorithms [6], percolation theory can help identify the critical road segments that have a significant impact on the formation of biking islands.

Compared to existing studies, our highlights rely on four aspects, in terms of the research focus, concept development, method, data as well as potential application.

First, existing studies are usually more focused on the analysis of general spatiotemporal patterns of bike usage (e.g., spatial distribution of trips across a week in a city), while this chapter pays particular attention to the investigation of geographical areas with a high concentration of bike usage (e.g., how to define the concept to describe them and how to quantitatively identify them). Second, this chapter proposes a new concept that can be used to delineate geographical areas of interest with a high concentration of bike usage; to the best of our knowledge, other studies have not attempted this kind of research effort to the concept development similar to biking islands, or to identify and analyse the underlying spatial structures similar to biking islands. Third, the research goals are achieved by using different methods: this paper adopts the percolation theory to identify biking islands for its suitability in describing the formation of clusters and critical road segments that have a significant influence on biking behaviour in an urban context [2, 11, 14]. Fourth, one of the best bike trip datasets is used to support this analysis. This dataset, provided by a market-leading company (Mobike), contains a large number of real GPS-enabled bike trajectories (about 0.72 million trips). Existing studies either used a much smaller trajectory dataset [4], a dataset self-collected by the researchers [25] or a dataset that only contains the origins and destinations of bike trips [7]. And finally, our analysis is not only helpful to understanding the travel behaviour of cyclists and urban structures used for cycling, but also has multifaceted implications and can support relevant urban and transport planning, especially about how to plan designated non-motorised areas for cycling and biking facilities and how to identify critical road segments that can improve the cycling efficiency of the entire network.

1.2 Study Area and Data

1.2.1 Study Area

Shanghai, with a population of 24 million and spanning an area of 6,341 square km, stands as one of the largest and most dynamic cities in the world. As of July 2017, approximately 1.5 million dockless shared bicycles, operated by various service providers, were in circulation across the city. Administratively, Shanghai is divided into 16 districts. However, for the purpose of this study, the Chongming District is excluded from the analysis due to its notably low density of bike usage recorded in our dataset. Consequently, the remaining 15 districts are selected as the study area, as highlighted in Fig. 1.1. The central urban area of Shanghai is generally defined by the region enclosed within the outer ring road. A prominent geographical feature within this central area is the Huangpu River, which serves as a natural boundary, dividing the city into two distinct sections: *Pudong*, referring to the eastern side of the River, and *Puxi*, denoting the western side. This division not only reflects historical and economic contrasts but also plays a crucial role in shaping cycling patterns and mobility behaviours within the city. The study focuses on capturing these spatial

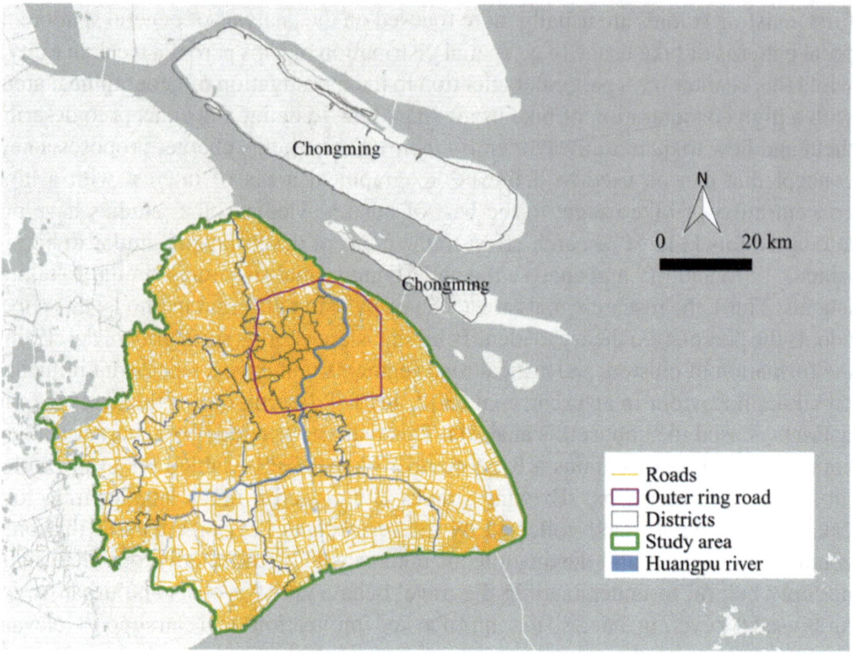

Fig. 1.1 Study area and road network

dynamics to better understand how dockless bike-sharing operates across Shanghai's diverse urban landscape.

1.2.2 Bike Trajectory Data

The bike trajectory data utilised in this study are provided by Mobike, China's recognised leader in dockless bike-sharing during the period of analysis. As of March 2017, Mobike operated approximately four million shared bicycles, collectively generating around twenty million trips per day. The trajectory dataset employed in this research is derived from a random selection of users, all of whom had registered their accounts with Shanghai as their specified location. Given that users registered in Shanghai may also undertake trips outside the city, journeys occurring beyond the defined study area are filtered out to maintain geographic relevance. Additionally, to ensure data quality and remove outliers, trips with abnormal characteristics are excluded—specifically, only trips with a length between 50 m and 50 km and a duration exceeding 60 s are retained. Following this data preprocessing, the final dataset comprises 720 thousand trips made by 130 thousand users, utilising 290 thousand bikes. These trips span a 15-day period from 16 to 30th September 2017. Each record contains basic trip information, such as trip ID, start and end times of a trip, longitude and latitude of the origin and destination of a trip and *trajectory*. The attribute *trajectory* contains spatiotemporal information about each trip's trajectory. In other words, the attribute *trajectory* is a set of GPS points, for instance, $l1 \rightarrow l2 \rightarrow \cdots \rightarrow ln$, where each GPS point contains geospatial coordinates and a time stamp, i.e., $l = (longitude, latitude, timestamp)$.

1.2.3 Bike Road Network

This study employs OSMnx to extract Shanghai's bike road network. OSMnx is a Python package designed to download and analyse road networks from OpenStreetMap. Recognising the potential for missing bikeable roads in the extracted network, the data quality is manually validated by comparing the road network with a random sample of 50,000 raw trajectories. Specifically, for each sampling point, the distance between the GPS point and its nearest road segment was measured. Distances exceeding 50 m are flagged as potential indicators of missing roads. To further verify these discrepancies, Google Satellite images and Baidu Street View—a leading map service in China—are then used as references to verify missing roads. Following this validation process, an additional 10,700 road segments are identified and incorporated into the original network extracted from OpenStreetMap. The finalised bike road network comprises 74,800 nodes and 106,300 segments, as illustrated in Fig. 1.1.

1.3 Methods

The methodological framework adopted in this study is illustrated in Fig. 1.2. It outlines the step-by-step process, from data preparation and bike usage measurement to the application of percolation theory and the final identification of biking islands. Each component of this framework is elaborated upon in the following subsections, providing a comprehensive explanation of the techniques and analytical approaches employed.

1.3.1 Bike Trajectory Processing

The primary objective of trajectory processing is to deduce the actual travel routes for each bike trip, which subsequently serve as the basis for measuring bike usage. The raw dataset undergoes an initial cleaning process by excluding trips with abnormal durations or implausible trip lengths. Specifically, trips lasting less than 60 s or covering distances outside the range of 50 m–50 km—measured by the raw trajectory length—are regarded as outliers and are removed from the dataset. Upon further

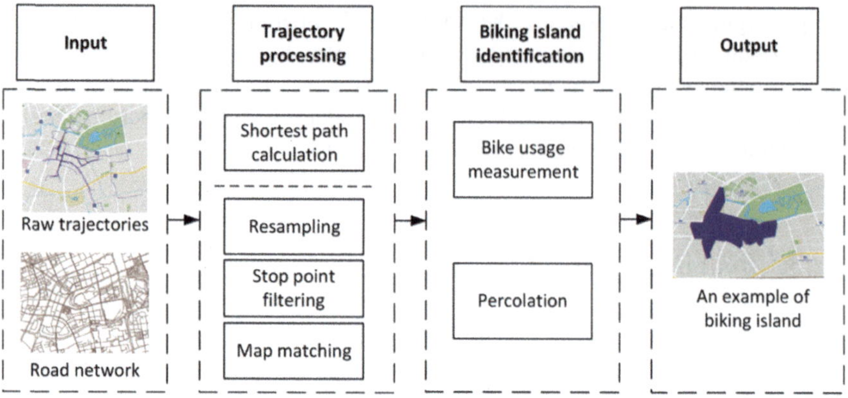

Fig. 1.2 The workflow of the biking island identification

examination of the remaining trajectories, it is observed that a small proportion of trips lack GPS sampling points, containing only the start and end locations. For these trips, the real paths are inferred using the shortest path algorithm. In contrast, the majority of trips include GPS sampling points and are characterised by a high sampling rate, with 54% of trips having a sampling interval of less than 5 s. However, these trajectories often contain noise, likely resulting from GPS inaccuracies or brief stops made by users during their journeys. To enhance the accuracy and reliability of the data, the trajectories with GPS sampling points undergo an additional three-step processing procedure, shown as follows:

Step 1: Resampling. A high sampling rate of trajectories offers greater details regarding users' trips. However, it can also generate a substantial number of redundant points and introduce additional errors during the trajectory map-matching process. To mitigate this issue, the trajectories are resampled using a relatively larger sampling interval. The selection of the resampling interval is carefully determined by considering both the original sampling rate and the users' movement speed. This approach aims to strike a balance between computational efficiency and the preservation of essential information necessary for accurate map matching.

Step 2: Stop point filtering. GPS trackers continuously collect positional data, even when a cyclist is stationary. Such stops can produce clusters of redundant GPS points—referred to as stop points—around a single location, potentially introducing errors during the map-matching process [24]. To enhance the accuracy of trajectory data, it is crucial to filter out these stop points prior to performing map matching. In this study, an orientation-based method is employed to identify and remove stop points from bike trajectories, following the approach outlined by Sultan et al. [24]. To be specific, for a trajectory with n points $\{p_1, p_2, ..p_n\}$, the distances between two consecutive points are represented as $D = \{d_1, d_2, \ldots d_{n-1}\}$, and $d_i = \text{Distance}\big((x_i, y_i), (x_{i+1}, y_{i+1})\big)$. Then, a search radius $r = \text{ mean } (D) + 2 * \text{ std } (D)$ can be defined for the trajectory, where mean (D) and std (D) represent the mean and standard deviation of D, respectively. For each sampling point, a set of candidate points located within r is constructed. The orientations between the examined point p and each candidate point p_i are measured by the angle (from 0 to360°) between the local north and the vector $\boldsymbol{v}_{pp_i}$. Each candidate point's orientation is then assigned to one of the eight evenly split slices. More than 4 different slices detected represent a lack of continuity in the user's cycling, the point is detected as a stop point.

Step 3: Map matching. To accurately identify the travel routes of bike users, it is essential to align the resampled GPS points with the road network—a process commonly addressed through map-matching algorithms [22]. For this study, the fast map matching (FMM) algorithm, developed by Yang and Gidófalvi [26], is selected due to its computational efficiency in matching large volumes of trajectory data to extensive road networks. The FMM algorithm comprises two key components: precomputation and map matching. For the precomputation, the algorithm iterates through each network node using a delta-upper bounded Dijkstra query to calculate the shortest paths between the current node and its neighbouring nodes. These precomputed shortest paths are then stored to accelerate the subsequent map-matching process. In terms of the map matching, the algorithm adopts a strategy similar to the hidden Markov model (HMM)-based approach proposed by Newson and Krumm [17], aiming to identify the optimal matching path from the set of candidate paths. The candidate paths are constructed through the following steps: (a) Identifying the k nearest road network edges for each sampling point within a specified distance threshold, with the projected point on each edge serving as a candidate matching point; (b) Retrieving the shortest path distance between each pair of neighbouring candidate matching points using the precomputed shortest paths; and (c) Constructing candidate paths by selecting one matching point for each sampling point from their respective candidate sets. To determine the most likely travel route, a score function is defined by combining the emission probabilities of the sampling points with the transition probabilities between consecutive points. The candidate path with the highest score is ultimately selected as the final matched route.

1.3.2 Identify Biking Islands Using Percolation Theory

Percolation theory has been extensively employed to examine the formation of clusters within random graphs and to explore how clusters of varying sizes emerge when a critical threshold probability P_c is reached [11]. Originally introduced by Broadbent and Hammersley [5], this theory has since been widely adopted by researchers to address a range of problems across natural and material sciences [1, 15]. Its applications have provided valuable insights into phenomena such as the spread of diseases, fluid flow through porous materials, and the resilience of complex networks. The flexibility and mathematical rigour of percolation theory make it a powerful tool for analysing connectivity and fragmentation patterns within diverse systems.

When a percolation transition occurs, percolation can act as a testing ground for examining critical phenomena and rules for system behaviour. In recent decades, the theory has also been increasingly applied in social sciences [11]. Solomon et al. [23] linked the presence of extreme market shares in the media industry—where shares approach either 0% or 100%—to a percolation process involving networks of diverse consumers. Kanai et al. [12] introduced a price model grounded in percolation theory to reproduce the price distribution of components used in certain industrial goods. Although relatively limited, there have been efforts to utilise percolation theory in the context of transport and spatial analysis. For instance, Li et al. [14] explored how local traffic flows on roads coalesce into broader, city-level traffic patterns. Their study, conducted in a central business district of Beijing, China, revealed that global traffic flow is dynamically formed by clusters of local flows interconnected through critical bottleneck links. Arcaute et al. [2] utilised percolation theory to reveal the hierarchical organisation of cities and regions across Britain at multiple spatial scales. Their study examined the evolving urban forms by calculating the fractal dimension of clusters at each distance threshold during the percolation process. Integrating network science with percolation theory, Mostafizi et al. [16] proposed a multi-agent approach to assess the impact of connected vehicle networks under different levels of market penetration. Furthermore, Jiang et al. [11] combined transport and economic factors to construct a network representing the Yangtze River Economic Belt in China, employing percolation theory to analyse the relocation patterns of the manufacturing industry within this network. Existing studies demonstrated the usability of percolation theory in identifying clusters (such as cities and local traffic flows) in an urban context. This study will explore its applicability for identifying biking islands using high-resolution trajectory data.

Percolation theory, dealing with the connected clusters of a random graph, is explained by using a weighted graph $G = (V, E)$ as an example, where G represents a graph, V and E represent its associated nodes and edges, respectively. The edge e_{ij} between nodes v_i and v_j is opened based on a probability $p_{e_{ij}}$ (closed with a probability of $1 - p_{e_{ij}}$), and the probability of each edge (i.e., the edge's weight) is assumed to be independent. Percolation theory is able to reveal the characteristics of clusters under different probability thresholds. To be specific, for a given threshold p_s (ranging from 0 to 1), the edges can be classified as blocked or connected based on:

$$e_{ij} = \begin{cases} 1, p_{e_{ij}} \geq p_s \\ 0, p_{e_{ij}} < p_s \end{cases} \tag{1.1}$$

where, $e_{ij} = 1$ means node i and j are connected when the probability is larger than p_s; otherwise, the nodes are considered to be disconnected. Given a threshold, the states of all edges (i.e., connected or blocked) can be determined based on Eq. (1.1). By finding the connected components, clusters under this threshold can then be formed and obtained. Then, a cluster's size is measured by the number of nodes contained in this cluster [2]. In this manner, by setting different thresholds, the entire graph exhibits different states and may split into various sub-connected components (i.e., clusters).

Figure 1.3 gives an illustrative example, where the probabilities of the edges are shown in Fig. 1.3a. When $p_s = 0$, the graph represents the original road network. When setting p_s as 0.3, 0.5, and 0.7, respectively, the corresponding clusters are shown in Fig. 1.3b–d, respectively. As illustrated, by increasing the value of p_s, the overall number of connected nodes decreases, and the number of clusters changes as well. When $p_s = 1$, the entire network is blocked without any cluster. As p_s increases, the size of the largest component starts to decrease, and the size of the second largest cluster begins to increase. According to percolation theory, at the point when the second largest component reaches its maximum size, the network is transitioned from a connected phase to a fragmented phase [14]. This transition point is referred to as the critical point and denoted as p_c. Edges blocked at p_c plays a key role that bridges different functional clusters and can be considered as bottlenecks, because their weights (i.e., probabilities) are the lowest with respect to the other edges that are connected under this point (i.e., critical point) and p_c is determined by their weights. For example, when increasing the threshold from 0 to 1 with an interval of 0.1, the example network in Fig. 1.3 reaches its critical point at the threshold of 0.5, and the bottleneck edge is marked by the red circle in Fig. 1.3c.

In the context of the biking island identification, the objective is to detect clusters with a high density of bike usage. The road network is conceptualised as a graph, where nodes represent intersections and edges correspond to road segments. As previously outlined, bike usage is quantified by counting the number of trips along each road segment. The weight of an edge is determined by calculating the ratio

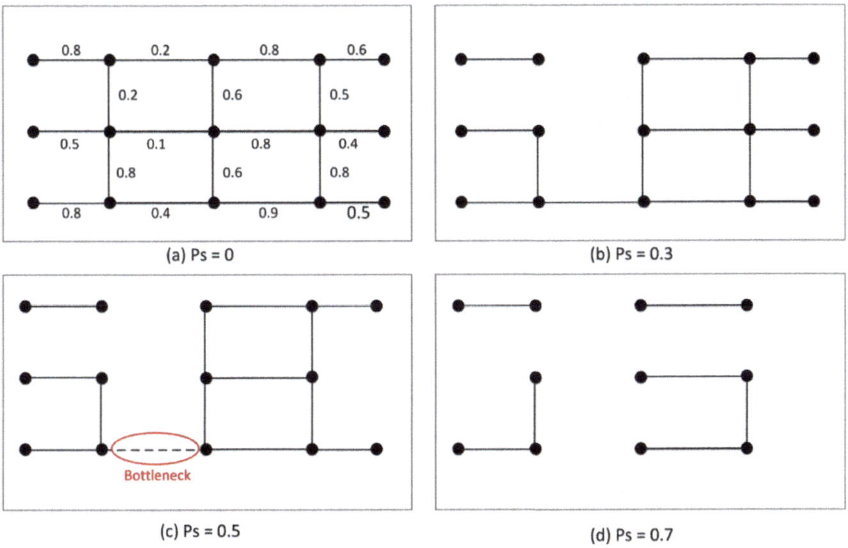

Fig. 1.3 An illustrative example of a percolation process

between the number of trips on that edge and the maximum trip count observed across all edges—essentially normalising the values by the maximum. For a given probability threshold, edges with weights below the threshold are considered disconnected (assigned a value of 0), while those above the threshold are marked as connected (assigned a value of 1). By systematically adjusting the threshold value, it becomes possible to identify biking islands of varying intensity in terms of bike usage. The critical point within this framework emerges when the size of the second-largest cluster reaches its peak, signalling a pivotal moment in the network's connectivity. At this stage, bottleneck edges can be pinpointed by identifying the edges that become disconnected. Drawing a parallel to critical transport links that influence traffic efficiency [14], these bottleneck edges represent the key road segments whose disconnection would disrupt the continuity of biking across the broader road network.

1.3.3 Percolation Theory and DBSCAN Algorithm

This sub-section aims to illustrate the connection between percolation theory and the Density-Based Spatial Clustering of Applications with Noise (DBSCAN) algorithm in terms of the biking island identification. Specifically, it is possible to obtain the same biking islands using these two methods. With respect to the percolation theory-based biking island identification, the input for percolation theory is a road network, where each edge e_{ij} with a weight equal to its trip number (denoted as $t_{e_{ij}}$). Given a threshold T, the corresponding biking islands can be found by following the procedures described in Sect. 1.3.2.

The DBSCAN algorithm is a well-established clustering method, primarily designed for point clustering. It classifies each point either as part of a cluster or as noise, depending on the density distribution of the input points. The algorithm relies on two key parameters: *eps* and *minPts*. The parameter *eps* defines the maximum distance between two points for one to be considered within the neighbourhood of the other, while *minPts* specifies the minimum number of points within a point's neighbourhood required for it to be classified as a core point. In the context of the biking island identification, the input points for the DBSCAN algorithm correspond to the nodes of the road network. To ensure the DBSCAN algorithm produces biking islands consistent with those derived from percolation theory, four essential steps must be followed in configuring the algorithm.

Step 1: Constructing the weight matrix. For running the DBSCAN algorithm, a weight matrix (denoted as W) is constructed for the input network nodes. Given two nodes n_i and n_j, if $\{n\}_{\{i\}}$ directly connects n_j *by* e_{ij} and $t_{eij} \neq 0$, then *the* weight w_{ij} between these two nodes is formulated as:

$$w_{ij} = \frac{t_{max}}{t_{eij}} \tag{1.2}$$

Otherwise:

$$w_{ij} = t_{max} \tag{1.3}$$

where, t_{max} is the maximum value of the trip number (i.e., $t_{e_{ij}}$) of the network edges. By following the transform of formula, the weight between two nodes is smaller when they are directly connected by an edge with a larger trip number. On the other hand, the weight of two nodes will be t_{max}, when they are indirectly connected or directly connected by an edge with a trip number equal to zero.

Step 2: Configuring the parameters of DBSCAN. Corresponding to a threshold T for percolation, the parameters for the DBSCAN algorithm are set as: $eps = \frac{t_{max}}{T}$; and $MinPts = 2$. In other words, $MinPts$ is fixed to be 2 for the algorithm, and the value of eps is depended on the corresponding threshold T. By fixing $MinPt$ to 2, any node with at least one node within the eps will be regarded as a core point during the clustering of the DBSCAN algorithm. Given $T \in (0, t_{max})$, it has $eps \in (0, t_{max})$. Then, two nodes are impossible to be classified into a cluster if they are indirectly connected or directly connected by an edge with a trip number equal to zero, because the weight between them is t_{max}.

Step 3: Clustering the network nodes. Based on the constructed weight matrix W and defined eps and $MinPts$, the network nodes can be easily clustered into different biking islands by applying the DBSCAN algorithm.

Step 4: Adding connection information for the nodes of biking islands. A biking island generated by percolation theory is represented by a set of geographically connected network edges, while a biking island generated by the DBSCAN algorithm is represented by a set of network nodes. There is no information on how the nodes are connected with each other in a biking island generated by the DBSCAN algorithm. Therefore, this step is used to add the connection information for network nodes within each biking island. This can be achieved by iterating every pair of nodes (n_i, n_j) of each biking island. If $w_{ij} \leq \frac{t_{max}}{T}$, then a connection bewteen n_i and n_j is added; otherwise, no connection should be added between n_i and n_j.

By following the above four-step DBSCAN method, it is able to generate the same biking islands as that generated by percolation theory. Figure 1.4 gives an example of biking island identification based on percolation theory and the above four-step DBSCAN method. Figure 1.4a shows the weighted road network for percolation, where the weight of an edge is represented by its trip number. The maximum trip number of this road network is 9 (i.e.,$t_{max} = 9$). Correspondingly, Figure 1.4b shows the weighted road network for DBSCAN. Figure 1.4c shows the obtained biking islands based on percolation theory at the threshold $T = 3$. The corresponding biking islands obtained by the four-step DBSCAN method (i.e., $eps = \frac{9}{3}$, $MinPts = 2$) are shown in Fig. 1.4d. It is clear that the results obtained by these two methods are the same. The results generated by percolation theory at the threshold $T = 5$ are shown in Fig. 1.4e. Using the corresponding paramters ($eps = \frac{9}{5}$,$MinPts = 2$) for the four-step DBSCAN method, it is also able to obtain the same results (Fig. 1.4f).

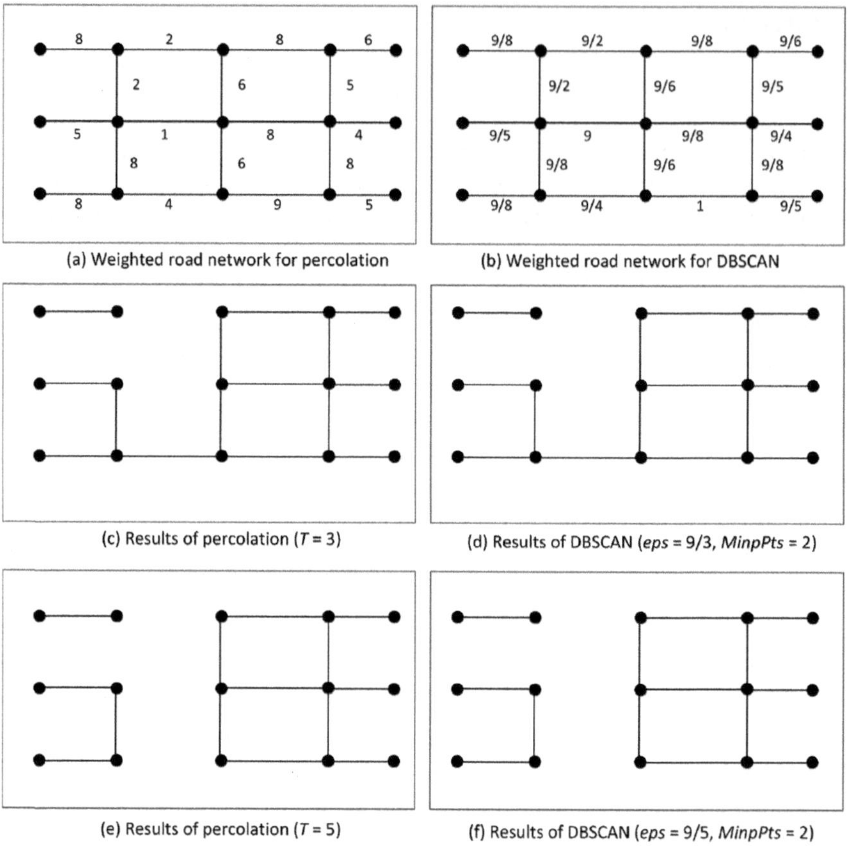

Fig. 1.4 An example of the biking island identification based on percolation theory and the four-step DBSCAN method

The above example demonstrated that it is possible to achieve identical results using both percolation theory and the DBSCAN algorithm, provided the four outlined steps are meticulously followed. However, the process of applying percolation theory is considerably more straightforward compared to the DBSCAN algorithm. More significantly, percolation theory offers a mathematically robust approach to identify critical road segments that hinder the continuity of biking across the network—an aspect that the DBSCAN algorithm does not inherently support.

1.4 Results

In the original trajectory dataset, 13.7% of the trips lack sampling points, and their routes are inferred using the shortest path algorithm. Figure 1.5 displays the statistics of sampling intervals for the remaining 86.3% of the trips. Notably, most trips have a sampling interval of less than 15 s. Based on the calculated average biking speed of 3.31 m/s, a resampling interval of 15 s is adopted, significantly reducing the number of GPS points from 92.2 million to 16.7 million.

During the stop point filtering process, 161 thousand trips are identified with stop points, leading to the removal of 794 thousand stop points. Ultimately, 15.9 million GPS points from 620 thousand trips are processed and aligned with the road network through the map-matching process. A sensitivity analysis is conducted to fine-tune the parameters for FMM: the pre-computation range for the shortest path algorithm (delta Δ) is set to 2.5 km, the search radius (r) to 50 m, and the number of neighbouring edges (k) to 6, and the GPS error (σ) to 30 m. By combining the estimated shortest paths for trips without GPS sampling and the matched paths for those with GPS data, the travel routes for all individual trips are reconstructed.

1.4.1 Characteristics of Bike Trips

The characteristics of bike trips are analysed from four different perspectives to uncover patterns in biking usage:

(1) Temporal distribution (Fig. 1.6a): Two peaks in bike usage are observed—7–9 am and 5–7 pm—corresponding to the morning and evening rush hours. This pattern suggests that dockless bike-sharing is predominantly used for commuting in Shanghai.
(2) Trip length distribution (Fig. 1.6b): The average trip length is 1,598 m. Most trips are within 2 km, indicating that bikes are primarily used for short-distance travel. The highest bike usage occurs in the 500–1,000 m range, with a notable decline

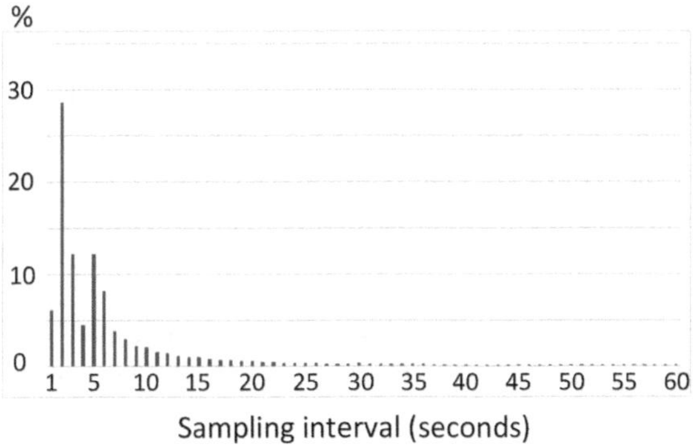

Fig. 1.5 The sampling interval distribution for trips with sampling points

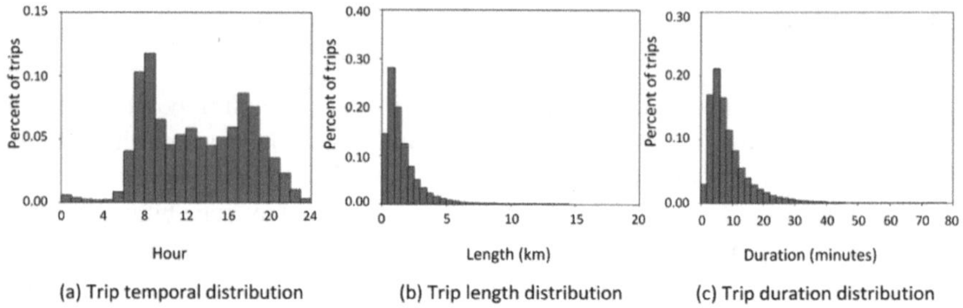

Fig. 1.6 Characteristics of bike trips

after 1,000 m. The long-tail pattern suggests that travel distance influences the willingness to bike.

(3) Trip duration (Fig. 1.6c): The average trip duration is 9.23 min. The pattern mirrors the trip length distribution, showing a sharp rise followed by a steady decline. Notably, 69% of trips are completed within 10 min, and 92% within 20 min.

(4) Spatial distribution (Fig. 1.7): The spatial distribution of trips across the road network is visualised by the number of trips on each road segment. The pattern is heterogeneous, with higher bike usage concentrated in the central urban area (notably in *Puxi*) and lower usage in suburban regions. Additionally, local hotspots are scattered across various areas.

1.4.2 General Analysis of Biking Islands

This chapter explores the evolution of biking islands by adjusting the probability threshold, which effectively reflects varying levels of bike usage. Given that the maximum number of bike trips recorded across all road segments remains constant, the probability thresholds can be directly translated into corresponding trip counts. The size of a biking island is quantified by the number of road junctions it encompasses. The statistical outcomes of the biking island identification under changing probability thresholds are presented in Fig. 1.8. Specifically, Fig. 1.8a illustrates the variation in both the total number of biking islands and the subset of islands containing more than 10 road junctions. Meanwhile, Fig. 1.8b shows the trend in the sizes of the largest and second-largest biking islands as the threshold changes. As depicted in Fig. 1.8a, both curves exhibit initial fluctuations, followed by a steady decline once a certain threshold is exceeded. The observed fluctuations suggest that some biking islands are prone to fragmentation along local road segments, likely due to the uneven distribution of the cycling demand across different areas. Additionally, the noticeable gap between the two curves indicates that the majority of identified biking islands are relatively small in scale. This outcome may be attributed to typical

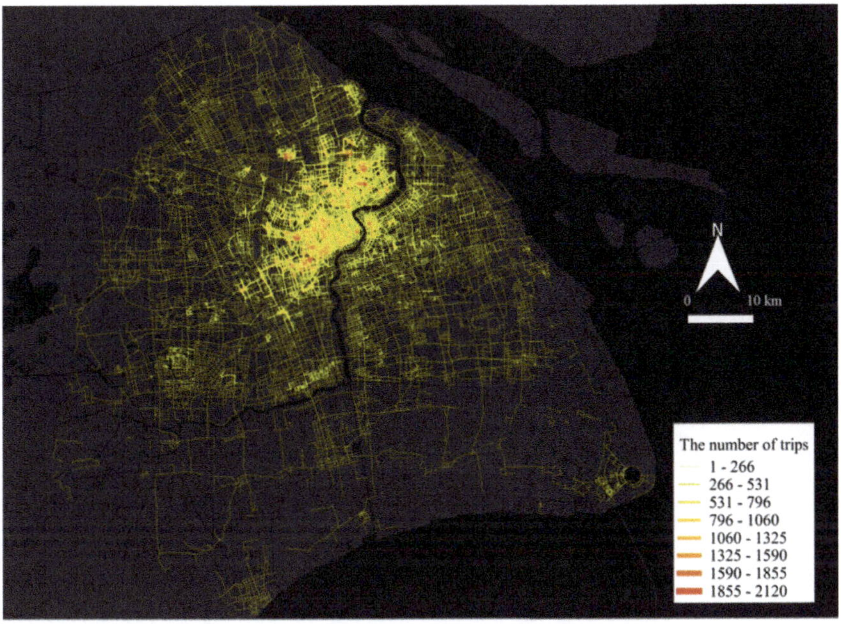

Fig. 1.7 The number of bike trips distributed within the road network

characteristics of bike trips, which generally involve short travel distances and brief durations.

To further analyse how the biking islands are distributed spatially and their characteristics, this study takes a closer look at 9 different thresholds: 50, 150, 250, 350, 450, 550, 650, 750, and 850. Figure 1.9 shows their respective distributions. For a better visualisation, only the top-10 largest biking islands are shown and the corresponding lengths of roads (*LRs*) and average trips numbers (*AVTs*) are given in Table 1.1. In general, the evolution of biking islands demonstrates a hierarchical structure (Fig. 1.9). Specifically, with the increase of threshold, the biking islands start to shrink and split into various smaller ones. For example, comparing Fig. 1.9a and b, it is obvious that the top-10 biking islands in Fig. 1.9b are all originated from the largest biking island in Fig. 1.9a. In terms of spatial distribution, most of the large-size biking islands are located in the central area of Shanghai, especially in *Puxi*, where high bike usage is observed. As indicated by the numerical evidence in Table 1.1, the sequential ID based on biking island size, biking island *LRs*, and *AVTs* do not follow a same trend with each other, which may be explained by the differences in density, connectivity, and length of road segments within different biking islands. For example, under the threshold of 750, the *LR* of the 4th largest biking island (4 km) is smaller than the *LR* of the 5th largest biking island (6.4 km). Meanwhile, in terms of *AVT*, the 7th has the largest value (1251), followed by the 6th (1082) and the 4th (1018) largest biking island. Therefore, when conducting relevant biking island planning, multiple factors should be taken into consideration.

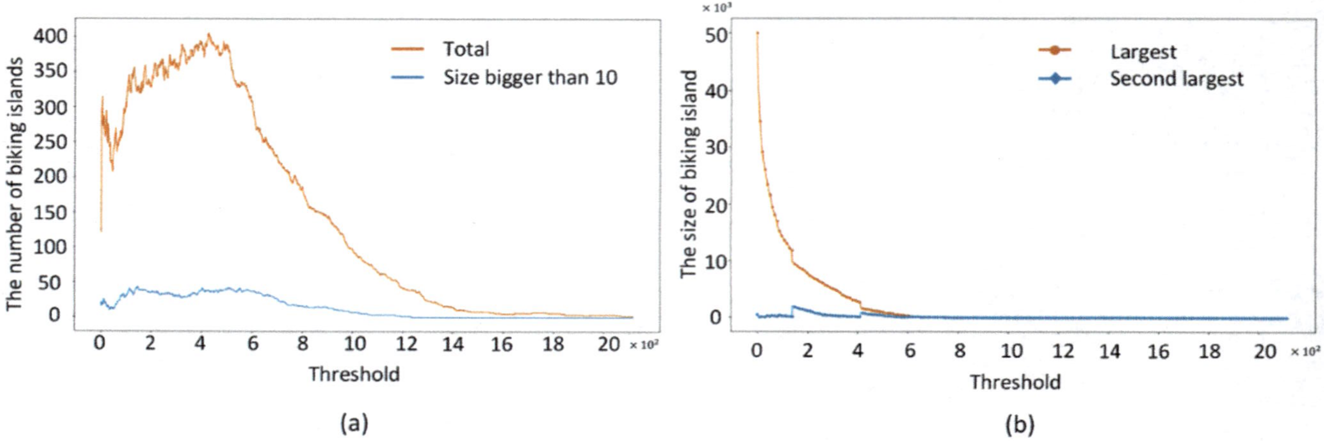

Fig. 1.8 Biking islands statistics under different thresholds. Note: (**a**) The number of biking islands; (**b**) The sizes of the largest and second-largest biking islands

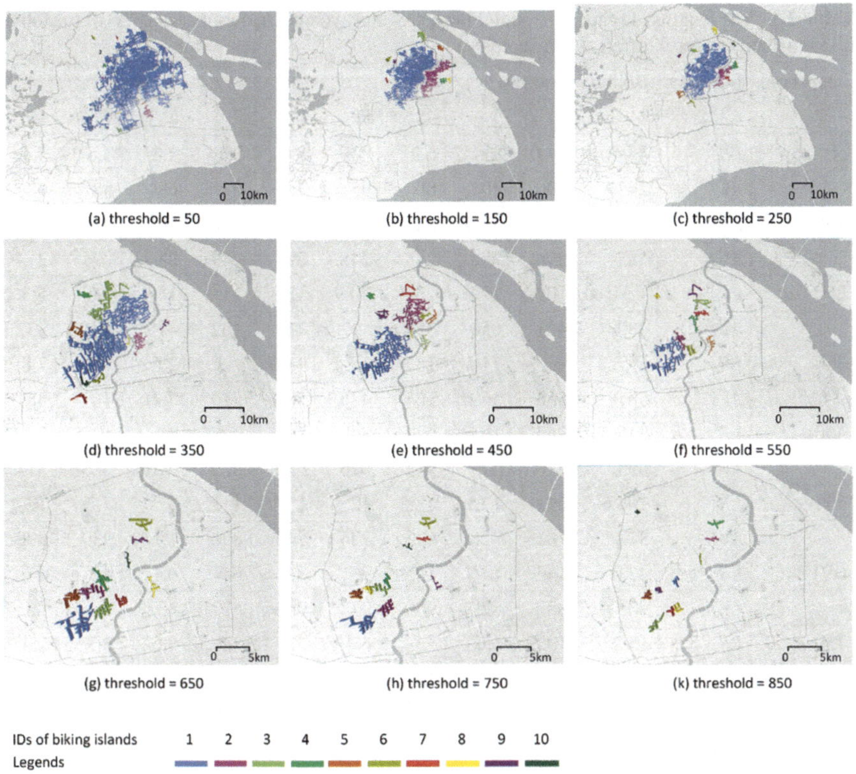

Fig. 1.9 Spatial distributions of the top-10 largest biking islands under 9 different thresholds

When the threshold is low, an extensive biking island emerges in the central area. As the threshold gradually increases, this large island progressively fragments into two distinct biking islands. This transition aligns with the formation of the largest and second-largest clusters in percolation theory, reflecting the shift of the graph from a connected phase to a fragmented phase during the percolation process. As shown in Fig. 1.8b, when $p_c = 141$, the second largest biking island reaches its maximum size and this threshold is therefore deemed as the critical point. The corresponding spatial distribution of the top-10 largest biking islands at this transitioning phase (i.e., right before and at the critical point p_c) is shown in Fig. 1.10. The bottleneck road segment is highlighted by a red dotted line in Fig. 1.10a, which is the ferry line of Yangfu. The findings highlight that the Huangpu River serves as a natural barrier, significantly hindering the connectivity between biking islands in *Puxi* and *Pudong*. The relatively small value of p_c suggests that few individuals choose to cycle across the river, which is understandable given the absence of dedicated bike lanes. Cyclists are required to use the ferry service to transport their bikes across the river, adding to the inconvenience. To verify that the identification of the bottleneck edge—specifically, the Yangfu ferry line—at the critical point is not a random occurrence

Table 1.1 The lengths of roads (*LRs*, unit kilometre) and average trip numbers (*AVTs*) of the top-10 biking islands under 9 different thresholds

Threshold	Indicator	Biking island ID									
		1	2	3	4	5	6	7	8	9	10
50	*LR*	4,960.9	44.0	8.6	6.2	3.8	5.0	3.1	2.5	3.6	3.4
	AVT	242	100	108	86	95	78	67	68	58	48
150	*LR*	1,947.5	376.9	42.9	24.9	18.7	14.4	6.3	10.8	8.4	8.0
	AVT	387	296	288	248	248	297	249	286	291	274
250	*LR*	1,210.0	109.8	17.8	23.7	20.9	12.8	11.9	15.9	7.5	5.3
	AVT	480	421	650	354	412	417	402	415	381	332
350	*LR*	657.2	36.7	71.4	12.6	15.4	11.0	7.3	2.6	7.2	7.8
	AVT	581	556	569	765	486	581	510	750	461	481
450	*LR*	245.5	96.8	16.6	6.7	8.8	8.5	9.4	2.3	12.4	2.1
	AVT	695	710	675	911	675	630	829	616	628	514
550	*LR*	106.5	19.6	17.1	12.5	12.3	13.1	11.5	3.5	7.2	12.4
	AVT	813	781	826	823	724	745	890	1,112	939	898
650	*LR*	27.0	14.9	14.7	10.3	9.7	8.9	6.5	6.1	5.1	4.0
	AVT	903	887	879	900	937	966	829	832	1,105	885
750	*LR*	15.2	10.5	6.6	4.0	6.4	5.8	3.2	4.6	3.7	2.4
	AVT	970	970	968	1,018	996	1,082	1,251	947	915	980
850	*LR*	3.3	3.2	5.8	4.0	4.0	1.9	2.5	3.8	3.0	0.7
	AVT	1,040	1,251	1,157	1,157	1,138	1,028	1,044	1,036	1,026	1,561

[14], this study further examines the bottleneck edges during three different time periods (7:00–8:00, 12:00–13:00, and 17:00–18:00). The results consistently show that the Yangfu ferry line remains the bottleneck edge, restricting the connection between the largest and second-largest biking islands. This reinforces the crucial role of the Huangpu River in disrupting cycling continuity across Shanghai. Beyond the ferry line, the complex spaghetti junction created by the viaducts (Fig. 1.10b) also acts as a barrier, further limiting cycling connectivity between the second-largest and fourth-largest biking islands.

1.4.3 Selected Analysis of Biking Islands

To gain deeper insights into the formation of biking islands, it is crucial to explore why these clusters emerge in particular areas. This investigation is guided by two key questions: (1) which trips contribute to the formation of a biking island, and (2) why are these trips concentrated in certain areas? Regarding the first question, trips that pass through at least one road segment within a biking island are identified and classified as associated trips for that island. To better understand spatial patterns of these

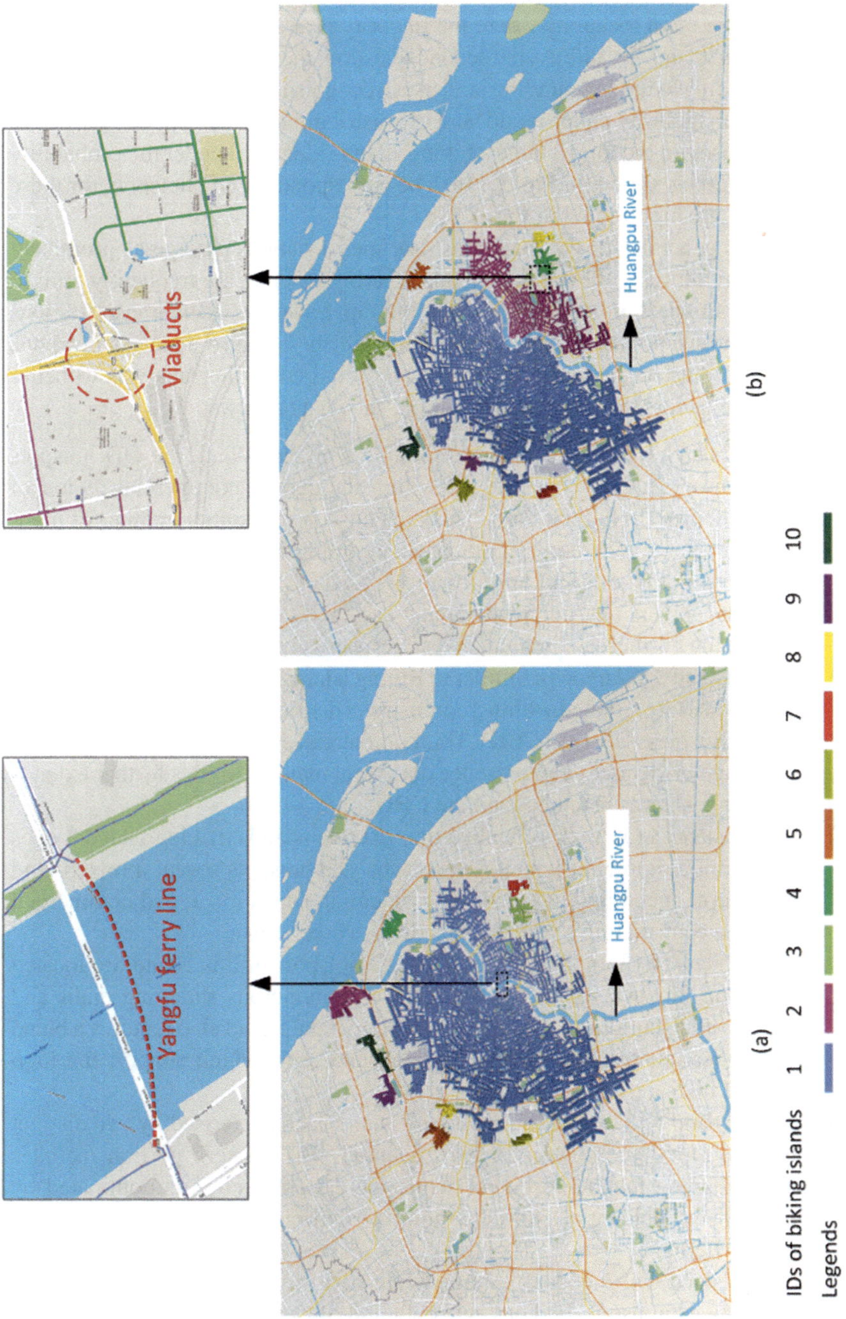

Fig. 1.10 Biking islands when the threshold is 140 (**a**) and 141 (**b**) (the critical point)

trips, the origins and destinations (ODs) of the associated trips are analysed using a kernel density function. This approach allows for the detection of OD hotspots, highlighting the main locations where the majority of associated trips begin or end for each biking island. To address the second question, the identified OD hotspots are labelled according to the surrounding land uses, recognising the well-established link between human mobility patterns and land use distribution. For illustrative purposes, this study focuses on the top 10 largest biking islands identified under the probability threshold of 750, as shown in Fig. 1.11a. The corresponding OD hotspots for these biking islands are highlighted in Fig. 1.11b.

The results generally reveal that areas near metro stations emerge as prominent OD hotspots, reinforcing the notion that bike-sharing schemes are frequently utilised to tackle the first/last mile problem—facilitating travel between transit stations and users' starting points or destinations. Beyond these metro-adjacent hotspots, additional OD clusters have been detected and required further analysis to better understand the factors driving their formation, shown as follows:

(1) Biking Island 1 (75 nodes): It is the largest biking island, the OD hotspots of which are mainly concentrated within the central commercial region of the Caohejing High-Tech Park. This area hosts a substantial number of high-tech companies, generating significant commuting demand, which explains the high density of bike usage. Additionally, several OD hotspots are observed near hospitals, universities, and residential areas, facilitating the island's extension towards the northeast, south, and west, respectively.
(2) Biking Island 2 (52 nodes): Similar to Biking Island 1, the primary OD hotspots of this island are situated within a commercial district, specifically the Xujiahui Central Business District (CBD). This commercial concentration largely drives bike usage in the area. The presence of several major hospitals further extends the biking island's reach beyond the CBD.
(3) Biking Island 3 (49 nodes): Unlike the first two islands, Biking Island 3 does not exhibit a dominant land use type. The OD hotspots reveal a mixed land use pattern, encompassing medical, residential, educational, and commercial facilities, contributing to the island's diverse bike usage.
(4) Biking Island 4 (36 nodes): The main OD hotspots of this island are located within the Jingan CBD, a bustling area with numerous shopping malls and modern office buildings. The eastern section of the island also features a blend of commercial, residential, and medical land uses, which support the high levels of bike activity.
(5) Biking Island 5 (34 nodes): This biking island is characterised by two distinct land use types—commercial and residential. The western portion includes large shopping malls and hotels, while the eastern section is dominated by high-density residential buildings, jointly contributing to the observed bike usage patterns.
(6) Biking Island 6 (33 nodes): A significant part of this island lies within two university campuses—Fudan University and Shanghai University of Finance and Economics—indicating that student travel plays a crucial role in shaping

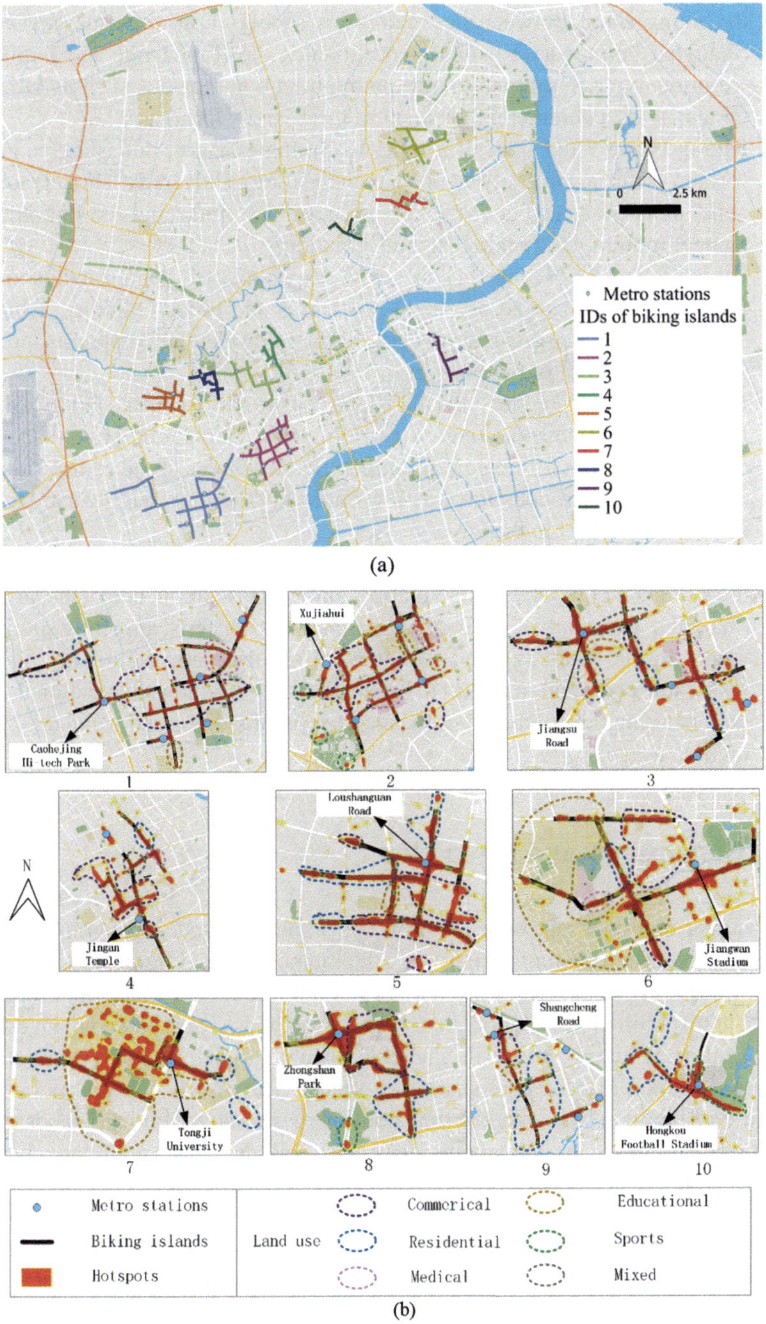

Fig. 1.11 Top-10 biking islands under the threshold of 750

the biking island. Between these campuses, a local OD hotspot emerges near the Shanghai Pulmonary Hospital. Additionally, the Knowledge and Innovation Community (KIC), home to various high-tech companies such as IBM and Oracle, adds a commercial element to the island's formation.

(7) Biking Island 7 (31 nodes): Similar to Biking Island 6, this island is largely formed within a university campus, namely Tongji University. The surrounding residential areas also contribute to bike usage, reinforcing the link between student mobility and biking island formation.

(8) Biking Island 8 (29 nodes): This island displays a clear land use division: the northern part is marked by shopping centres and office buildings, while the southern section consists of modern high-rise residential complexes, creating a balanced pattern of bike usage.

(9) Biking Island 9 (27 nodes): Biking Island 9 is primarily situated along Nanquan Road. The OD hotspots trace a corridor with commercial areas—shopping centres and hotels—to the north, three middle schools at the centre, and residential neighbourhoods towards the south, highlighting the diverse land use driving bike travel in the area.

(10) Biking Island 10 (22 nodes): The OD hotspots for this island cluster are around the Hongkou Football Stadium metro station. The stadium, one of the largest in Shanghai, frequently hosts football matches and concerts, attracting significant visitor traffic. The biking island extends from the metro station northwest to residential areas and southeast to Luxun Park, a well-known recreational space, underscoring the combined influence of sports facilities and residential zones on bike usage.

1.5 Conclusions and Discussion

This chapter introduces the concept of biking islands, defined as geographical areas with a high concentration of bike usage. The methodological framework for identifying these biking islands comprises two key components: bike trajectory processing and the application of percolation theory. These methods are employed to investigate the formation of biking islands in Shanghai, using a trajectory dataset provided by Mobike, a market-leading bike-sharing company. The results reveal a hierarchical structure of biking islands. As the threshold increases, larger biking islands gradually shrink and fragment into smaller ones. Most biking islands are relatively small in size, while the larger ones are predominantly situated within Shanghai's central urban areas. Notably, the Huangpu River serves as a natural barrier, disrupting the continuity of biking networks across the city. Furthermore, the formation of biking islands is closely linked to the surrounding urban functions. For instance, areas near metro stations or university campuses are more likely to form biking islands due to the high demand for short-distance travel.

The identification of biking islands allows for a clear distinction between areas with high and low concentrations of bike usage. This concept not only enhances our understanding of cyclists' travel behaviours and the spatial structure of urban cycling networks but also carries important implications for urban and transport planning. It can inform strategies for designating non-motorised zones, developing cycling infrastructure, and identifying critical road segments that could enhance the overall efficiency of the cycling network. Building on this concept, the study proposes that geographic areas of high bike usage can be conceptualised at multiple levels: a frequently used bike station can be seen as a *point of interest*, a popular bike lane as a *line of interest*, and a biking island as a *polygon or area of interest*. Effective transport planning must integrate these different spatial scales to develop a cohesive blueprint, particularly for non-motorised and active transport systems.

In this chapter, a biking island consists of a cluster of road segments and junctions within a road network. It can also be represented spatially by the area enclosing these clusters. For example, Fig. 1.12 illustrates a biking island using its convex hull, providing a clear visualisation of this area of interest. Additionally, since biking islands are identified through data-driven methods, a manual review of the road structure within these areas can offer further insights for planning. When constructing new bike lanes within a biking island, it is crucial to address any "short gaps" between road segments along the same route—such as those marked by red circles in Fig. 1.12—to maintain the continuity of the cycling network.

The selection of an appropriate threshold in percolation theory remains context-dependent. A lower threshold tends to produce larger biking islands, whereas a higher threshold results in fewer, more compact biking islands with higher levels of bike usage. In practical applications, identifying biking islands through percolation

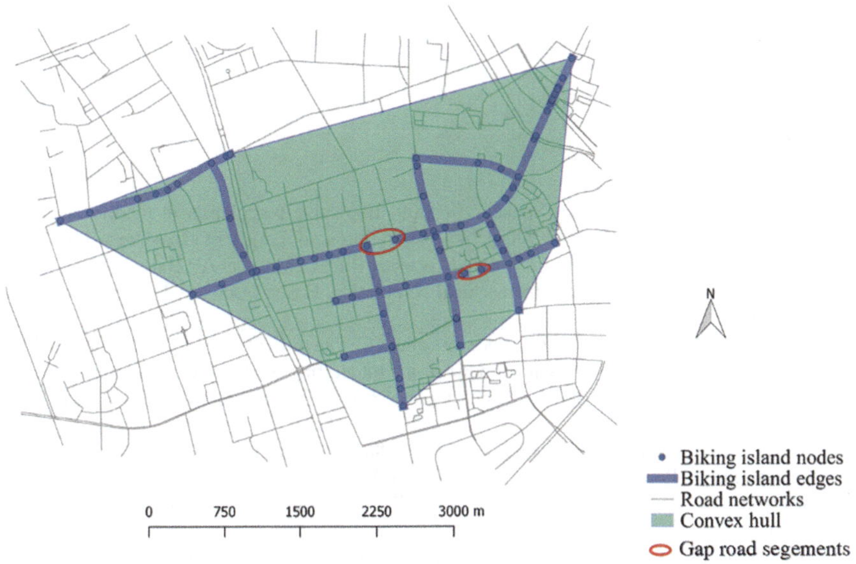

Fig. 1.12 An example of the convex hull of a biking island

theory could benefit from the development of an interactive interface. Such a tool would enable planners and decision-makers to explore the evolution of biking islands under different thresholds, allowing them to select suitable values based on specific planning goals. Furthermore, directly applying percolation theory to the entire road network may overlook local clusters due to spatial heterogeneity of bike usage. To address this, the network can be divided into sub-regions—for example, based on administrative boundaries or patterns of bike use—before applying percolation theory within each sub-region. This approach helps capture local variations and provides a more nuanced understanding of the biking island formation.

Building upon the findings of this study, several avenues for future research can be explored. The first is the classification of biking islands. It would be valuable to classify biking islands based on exogenous factors such as users' socioeconomic status, weather conditions, and other environmental variables. Different categories of biking islands may exhibit distinct characteristics, providing deeper insights into cycling behaviour across various user groups and urban settings. The second one is the localised application of percolation theory. While this chapter applied percolation theory to identify biking islands across the entire city of Shanghai, the method could be extended to examine specific local areas of interest. This would enable the identification of local biking islands and bottleneck road segments within particular neighbourhoods or districts, offering more granular insights for targeted urban planning. The final aspect is the planning and design within biking islands. Another promising direction involves investigating the planning and design of bike lanes within identified biking islands. An interesting research question would be how to develop automatic or semi-automatic methods for designing continuous and efficient cycling networks within these areas, ensuring connectivity and accessibility for cyclists.

References

1. Achlioptas D, D'Souza RM, Spencer J (2009) Explosive percolation in random networks. Science (New York, N.Y.) 323(5920):1453–1455. https://doi.org/10.1126/science.1167782
2. Arcaute E, Molinero C, Hatna E, Murcio R, Vargas-Ruiz C, Masucci AP, Batty M (2016) Cities and regions in Britain through hierarchical percolation. R Soc Open Sci 3(4):150691. https://doi.org/10.1098/rsos.150691
3. Arranz-López A, Soria-Lara JA, López-Escolano C, Pueyo Campos Á (2017) Retail mobility environments: a methodological framework for integrating retail activity and non-motorised accessibility in Zaragoza, Spain. J Transp Geogr 58:92–103. https://doi.org/10.1016/j.jtrangeo.2016.11.010
4. Bao J, He T, Ruan S, Li Y, Zheng Y (2017) Planning bike lanes based on sharing-bikes' trajectories. In: Proceedings of the 23rd ACM SIGKDD international conference on knowledge discovery and data mining - KDD '17, pp 1377–1386. https://doi.org/10.1145/3097983.3098056
5. Broadbent SR, Hammersley JM (1957) Percolation processes. Math Proc Camb Philos Soc 53(03):629. https://doi.org/10.1017/S0305004100032680

6. Croitoru A, Wayant N, Crooks A, Radzikowski J, Stefanidis A (2015) Linking cyber and physical spaces through community detection and clustering in social media feeds. Comput Environ Urban Syst 53:47–64. https://doi.org/10.1016/J.COMPENVURBSYS.2014.11.002
7. Du Y, Deng F, Liao F (2019) A model framework for discovering the spatio-temporal usage patterns of public free-floating bike-sharing system. Transp Res C Emerg Technol 103:39–55. https://doi.org/10.1016/J.TRC.2019.04.006
8. Faghih-Imani A, Eluru N, El-Geneidy AM, Rabbat M, Haq U (2014) How land-use and urban form impact bicycle flows: evidence from the bicycle-sharing system (BIXI) in Montreal. J Transp Geogr 41:306–314. https://doi.org/10.1016/j.jtrangeo.2014.01.013
9. Forma IA, Raviv T, Tzur M (2015) A 3-step math heuristic for the static repositioning problem in bike-sharing systems. Transp Res B Methodol 71:230–247. https://doi.org/10.1016/j.trb.2014.10.003
10. García-Palomares JC, Gutiérrez J, Latorre M (2012) Optimizing the location of stations in bike-sharing programs: a GIS approach. Appl Geogr 35(1–2):235–246. https://doi.org/10.1016/J.APGEOG.2012.07.002
11. Jiang Y, Timmermans HJP, Yu B (2018) Relocation of manufacturing industry from the perspective of transport accessibility—an application of percolation theory. Transp Policy 63:10–29. https://doi.org/10.1016/J.TRANPOL.2017.11.003
12. Kanai Y, Abe K, Seki Y (2015) Price percolation model. Physica A 427:226–233. https://doi.org/10.1016/J.PHYSA.2015.02.021
13. Lee J, Han J (2007) Trajectory clustering: a partition-and-group framework. In: Proceedings of the 2007 ACM SIGMOD international conference on management of data
14. Li D, Fu B, Wang Y, Lu G, Berezin Y, Stanley HE, Havlin S (2015) Percolation transition in dynamical traffic network with evolving critical bottlenecks. Proc Natl Acad Sci 112(3):669–672. https://doi.org/10.1073/pnas.1419185112
15. Majdandzic A, Podobnik B, Buldyrev SV, Kenett DY, Havlin S, Eugene Stanley H (2014) Spontaneous recovery in dynamical networks. Nat Phys 10(1):34–38. https://doi.org/10.1038/nphys2819
16. Mostafizi A, Dong S, Wang H (2017) Percolation phenomenon in connected vehicle network through a multi-agent approach: mobility benefits and market penetration. Transp Res C Emerg Technol 85:312–333. https://doi.org/10.1016/J.TRC.2017.09.013
17. Newson P, Krumm J (2009) Hidden Markov map matching through noise and sparseness. In: Proceedings of the 17th ACM SIGSPATIAL international conference on advances in geographic information systems - GIS '09. https://doi.org/10.1145/1653771.1653818
18. Padgham M (2012) Human movement is both diffusive and directed. PLoS ONE 7(5):e37754. https://doi.org/10.1371/journal.pone.0037754
19. Pal A, Zhang Y (2017) Free-floating bike sharing: solving real-life large-scale static rebalancing problems. Transp Res C Emerg Technol 80:92–116. https://doi.org/10.1016/J.TRC.2017.03.016
20. Park C, Sohn SY (2017) An optimization approach for the placement of bicycle-sharing stations to reduce short car trips: an application to the city of Seoul. Transp Res A Policy Pract 105:154–166. https://doi.org/10.1016/J.TRA.2017.08.019
21. Pfrommer J, Warrington J, Schildbach G, Morari M (2014) Dynamic vehicle redistribution and online price incentives in shared mobility systems. IEEE Trans Intell Transp Syst 15(4):1567–1578. https://doi.org/10.1109/TITS.2014.2303986
22. Quddus MA, Ochieng WY, Noland RB (2007) Current map-matching algorithms for transport applications: state-of-the art and future research directions. Transp Res C Emerg Technol 15(5):312–328. https://doi.org/10.1016/j.trc.2007.05.002
23. Solomon S, Weisbuch G, de Arcangelis L, Jan N, Stauffer D (2000) Social percolation models. Physica A Stat Mech Appl 277(1–2):239–247. https://doi.org/10.1016/S0378-4371(99)00543-9
24. Sultan J, Ben-Haim G, Haunert JH, Dalyot S (2017) Extracting spatial patterns in bicycle routes from crowdsourced data. Trans GIS 21(6):1321–1340. https://doi.org/10.1111/tgis.12280

25. Xu Y, Chen D, Zhang X, Tu W, Chen Y, Shen Y, Ratti C (2019) Unravel the landscape and pulses of cycling activities from a dockless bike-sharing system. Comput Environ Urban Syst 75:184–203. https://doi.org/10.1016/J.COMPENVURBSYS.2019.02.002
26. Yang C, Gidófalvi G (2018) Fast map matching, an algorithm integrating hidden Markov model with precomputation. Int J Geogr Inf Sci. https://doi.org/10.1080/13658816.2017.1400548
27. Zhang Y, Lin D, Mi Z (2019) Electric fence planning for dockless bike-sharing services. J Clean Prod 206:383–393. https://doi.org/10.1016/J.JCLEPRO.2018.09.215
28. Zhang Y, Liu L (2018) Understanding temporal pattern of human activities using temporal areas of interest. Appl Geogr 94:95–106. https://doi.org/10.1016/J.APGEOG.2018.03.009
29. Zhang Y, Mi Z (2018) Environmental benefits of bike sharing: a big data-based analysis. Appl Energy 220:296–301. https://doi.org/10.1016/j.apenergy.2018.03.101

Chapter 2
Examine Cross-City Travel Behaviour and Its Associated Cycling Spatial Structures

2.1 Introduction

As a green and healthy mode of travel, bike-sharing is a form of micromobility rental service designed for short trips. It offers considerable convenience for travellers and is often regarded as an effective solution to the first-and-last-mile problem—a situation where there is poor connectivity between trip origins or destinations and public transport stations [24, 27, 46]. Due to its association with various socioeconomic and environmental benefits, bike-sharing is becoming increasingly popular. By 2022, approximately three thousand cities worldwide had implemented bike-sharing systems [31]. According to the data released by Research and Markets, the bike-sharing market was valued at 3 billion US dollars in 2020 and is projected to grow to 4 billion by 2026.

Bike-sharing companies typically operate in multiple cities nationwide, and in some cases, globally. This means that once a user purchases a company's membership, they have the possibility to access its services across different cities/countries. For example, Hellobike's monthly subscription (35 Yuan per month, approximately $4.89) was valid in more than 400 cities across China in 2020. As a result, a subscriber might use a bike to commute to work in their home city during the week and then enjoy a leisurely ride along a scenic route in another city at the weekend. Existing studies have shown that inner-city bike-sharing trips effectively capture the spatiotemporal characteristics of cyclists' mobility, revealing cycling spatial structures within cities at a micro-scale [5]. Similarly, cross-city trips—undertaken by travellers who use bike-sharing services in two or more cities—have the potential to uncover spatial structures across cities at a macro scale. Surprisingly, however, cross-city travel within the bike-sharing context has received little attention in the existing literature. Research on the mobility patterns of cross-city travellers and the associated spatial structures remains largely absent. The main reason for this gap may be the limited availability of relevant data.

Y. Zhang, *Understanding Shared Micromobility in China Using Massive User-Generated Trip Data*, Urban Sustainability,
https://doi.org/10.1007/978-981-96-8492-2_2

To address this gap, this chapter aims to investigate the spatial structure of cross-city travel. It utilises extensive dockless bike-sharing data and focus on Zhejiang Province, China, as the study area. The analysis comprises two main parts. First, this chapter defines and identifies two types of cross-city trips and uses various indicators to describe the basic characteristics of cross-city travel. Then, it employs a network-based approach to uncover the spatial structures emerging from cross-city travel behaviour. This chapter contributes to the existing literature in the following ways. Firstly, to the best of our knowledge, it is the first study to examine cross-city bike-sharing travel behaviour and reveal the cycling spatial structure across cities at a macro scale. Secondly, the findings have important implications for the cross-city operation of bike-sharing services, an aspect that has not been fully explored in previous research.

2.2 Literature Review

Shared bikes have attracted considerable research attention in recent years, leading to a rapidly expanding body of literature. Existing studies explore a wide range of topics, including the analysis of usage patterns and their associations with socioeconomic and built environment factors [7, 37, 45], the relationships between bike-sharing and other modes of transport—such as bus and taxi services—in terms of substitution and integration [17, 32], travel demand prediction [21, 38], and system optimisation and planning [2, 4, 45]. Furthermore, since 2020, the impact of the COVID-19 pandemic on bike-sharing has become a widely discussed research topic [9, 12].

Many existing studies have focused on uncovering the characteristics of travel mobility, particularly from a spatio-temporal perspective. Hu et al. [13] identified significantly different usage patterns between membership groups and revealed diverse imbalances in bike supply and demand across stations in Colorado, the United States. By analysing GPS records from more than 110,000 dockless bike-sharing trips in Boston, the United States, Sadeghinasr et al. [36] found that, contrary to the assumptions of widely used bike routing algorithms, cyclists often utilise sidewalks, soft-surfaced paths, parking lots, and driveways. Some studies have also conducted comparative analyses of travel behaviour across different micromobility modes—such as shared e-bikes and e-scooters [28]—or between dock-based and dockless bike-sharing systems [25].

The most commonly used bike-sharing data are trip records containing origins and destinations (ODs), which are highly effective for describing cycling spatial structures by treating origins and destinations as nodes and trips as edges. As a result, cycling spatial structures have been extensively analysed in the existing literature. Some studies directly investigate urban structures through cycling patterns. For instance, Zhang et al. [45] introduced the concept of "biking islands", defined as geographical areas of interest with a high concentration of bike usage. Using bike trajectory data and percolation theory, they identified biking islands in Shanghai,

China. This concept not only describes urban structures but also holds the potential for supporting transport practices, such as planning designated non-motorised zones for cycling. Chen et al. [5] explored whether existing administrative divisions align with the spatial structures generated by cyclists. They constructed a spatial interaction network model based on dockless bike-sharing data in Nanjing, China, and delineated the borders of urban activity zones. Their findings revealed that, compared to administrative boundaries, activity zones more accurately reflect cycling mobility and urban structures, offering valuable insights for transport policies. Other studies have focused on mobility patterns—represented as flow structures—from a network perspective [42, 44], which also inherently capture cycling spatial structures. As a pioneering study, Zaltz Austwick et al. [44] applied community detection techniques to identify spatial network structures, uncovering well-connected subnetworks within London's bike-sharing system. Incorporating temporal variations, Zhang et al. [47] employed various network methods—such as degree measures and the PageRank algorithm—to reveal dynamic characteristics of dockless bike-sharing in Beijing, China. To support their network analysis, they divided the city into a uniform urban grid, treating grids with at least one trip as nodes and the trips between them as edges.

Two main limitations can be identified in existing studies. First, cross-city travel behaviour facilitated by dockless bike-sharing has not been adequately explored or discussed in the literature. Instead, the majority of studies focus on examining other types of behaviour across different demographic groups, such as by gender [3, 6], age [26, 40], educational status [39], and membership type [13, 16]. For example, Su et al. [39] found that participants, who are students from Beijing Jiaotong University (China), are 18.0% more likely to park their bikes in an orderly manner when receiving warning messages, and 25.2% more likely when offered a monetary incentive, demonstrating that behavioural incentives may be an effective strategy to address disorderly parking. Böcker et al. [3] revealed that bike-sharing is less accessible and less frequently used by women and older age groups, with gender biases evident through spatial inequalities that favour male-dominated central employment areas.

Secondly, cycling spatial structures have not been derived from cross-city travel behaviour and investigated at a cross-city scale. Most researchers have focused on a single city as their study area, with particular interest in several Tier-1 global cities, including frequently discussed locations such as New York City [6], London [12], Shanghai [2], Seoul [15], and Singapore [37]. A few studies have provided comparative analyses across multiple cities [18, 30]. However, these studies typically calculated quantitative features for each city individually and then compared them to identify similarities or differences, or to reveal general mobility patterns. Their underlying assumption is that bike-sharing behaviour in one city is unrelated to that in another city, and thus cross-city travel, as discussed in this chapter, is often assumed to be non-existent. For instance, O'Brien et al. [30] characterised 38 global bike-sharing systems using a set of descriptive indicators, such as system size and docking station capacity, to derive insights into sustainable transport systems. Kou and Cai [18] analysed bike-sharing trip data from eight cities in the United States. They found that trip duration and distance followed a lognormal distribution in larger cities (e.g., Chicago and New York), while in smaller cities, the distributions varied among Weibull, gamma, and lognormal, due to the restrictions on users' movements imposed by geographical boundaries.

2.3 Study Area and Data

2.3.1 Study Area

Zhejiang is a southeastern coastal province of China (Fig. 2.1), covering an area of 101,800 km^2 with a population of 65.40 million. It is one of China's most economically developed provinces, with a total gross domestic product (GDP) of ¥7.35 trillion (approximately $1.07 trillion) in 2021 [48]. Administratively, Zhejiang comprises 11 prefecture-level cities and 90 county-level (or district-level) cities. A prefecture-level city typically governs several districts and counties. For instance, Hangzhou, the provincial capital of Zhejiang, administers 10 districts (e.g., Shangcheng, Binjiang and Xihu) and three counties (Chunan, Jiande, and Tonglu). Generally, districts cover urban areas and are situated in the central part of a prefecture-level city, whereas counties encompass both urban and rural areas and are often geographically distant from the city centre. Since districts and counties hold the same administrative rank—the county level—in China, we refer to them collectively as county-level cities in this study. This chapter aims to explore the cycling spatial structure at both the prefecture-level and county-level city scales.

2.3.2 Data

The dockless bike-sharing trip data is provided by Hellobike, China's leading bike-sharing service provider. As of December 2021, the company operated dockless bike-sharing services in more than 400 cities across China, with 530 million registered users, generating a total mileage of 32.3 billion kilometres and an estimated reduction of 1.52 million tonnes of carbon emissions [11]. The original dataset comprises 13.11 million shared bike trips in Zhejiang Province, recorded over a two-week period from 14 to 27 September 2020. Each trip includes basic attributes such as trip ID, anonymised user ID, bike ID, start and end times, as well as the longitudes and latitudes of the start and end locations.

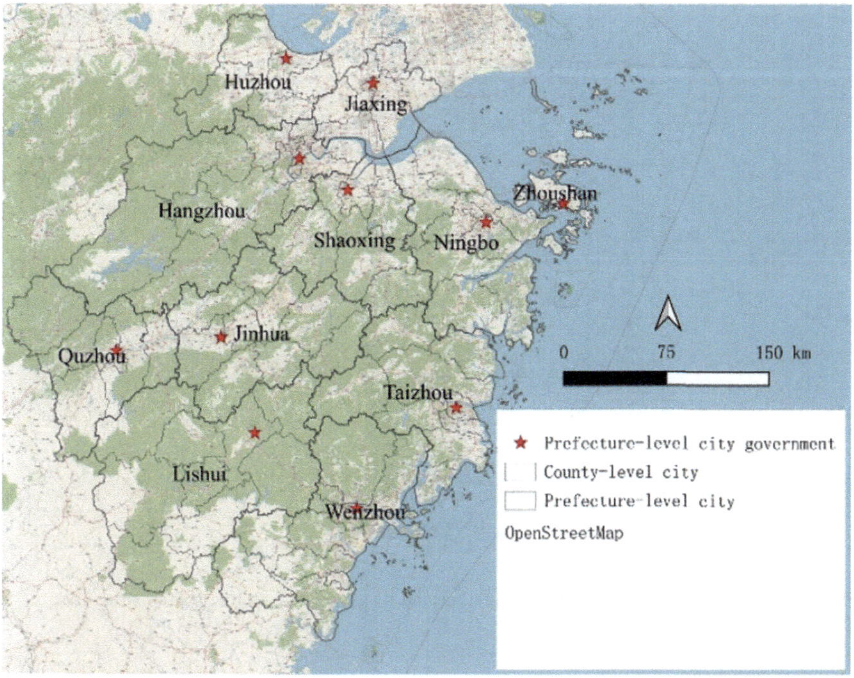

Fig. 2.1 The map of Zhejiang Province

2.4 Methodology

2.4.1 Define Cross-City Travel Behaviour

Cross-city travel behaviour refers to the movement of a traveller from one city to another. Such individuals are referred to as cross-city travellers, and their journeys spanning two cities are termed cross-city trips. This chapter focuses on cross-city travel within the context of bike-sharing. Broadly, two types of cross-city travel behaviour can be identified. The first type occurs when a rider travels from one city to another in a single trip—that is, the start and end locations of the trip are situated in two different cities. This is classified as a **Type I** trip, represented in green in Fig. 2.2a. The second type arises when a rider undertakes two consecutive inner-city trips in two different cities, implying the existence of cross-city travel between these cities. Although it is not possible to determine the exact travel mode used for the cross-city segment (e.g., bus, high-speed train, or even plane) solely from bike-sharing data, the data does confirm the occurrence of cross-city travel. In this case, the traveller indirectly generates a cross-city trip, classified as a **Type II** trip, shown in purple in Fig. 2.2a. For example, assuming there are four cities—*a*, *b*, *c*, and *d*—if a traveller completes two consecutive trips, *i* in City *a* and *k* in City *c*, we can infer the existence

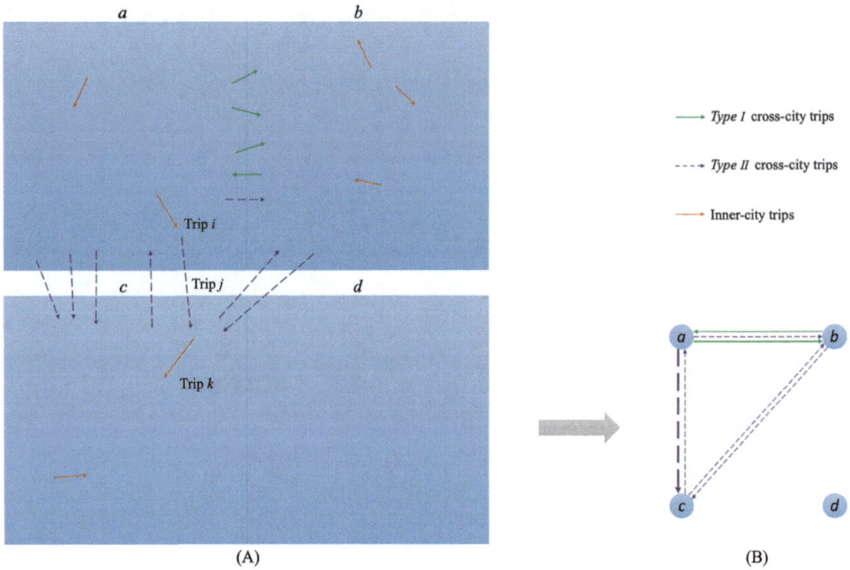

Fig. 2.2 Illustrating cross-city travel behaviour and its associated cycling spatial structure

of a potential trip *j*, representing the traveller's movement from City *a* to City *c*. In this instance, trip *j* is considered a **Type II** cross-city trip.

2.4.2 A Network-Based Approach to Reveal Cycling Spatial Structures

Graph theory is the study of graphs, which represent network structures through nodes and edges. Nodes can signify individuals or objects, while edges depict the connections or relationships between them. Network analysis, grounded in graph theory, is a method used to examine network structures and has been widely applied across physical and social sciences. From a network perspective, a cross-city trip can be conceptualised as an edge connecting two cities, represented as nodes within a network. The geographic centre of each city serves as a node, while a cross-city trip forms an edge between two nodes. The edge is weighted by the number of trips, with a higher trip count indicating more frequent spatial interactions between the two cities.

After identifying all cross-city trips, we can construct a network, which is referred to as the cycling spatial structure in this chapter. Figure 2.2b illustrates the cycling spatial structure derived from cross-city travel. The network consists of four nodes (representing cities) connected by six directed edges. The total number of trips is 12, which corresponds to a cumulative edge weight of 12. For example, there are four trips from City *a* to City *c*, so the edge between them has a weight of 4. In contrast, there is only one trip from *c* to *a*, giving that edge a weight of 1. Notably, City *d* is isolated from the other three cities, indicating no recorded cross-city trips involving City *d*.

To characterise the cycling spatial structure, we first employ several widely used indicators and methods to measure the network characteristics of the cycling spatial structure. These are outlined as follows:

- **Number of nodes**: This refers to the total number of nodes in a network. In our case, each node represents a city.
- **Number of edges**: This denotes the total number of edges in a network. An edge represents a connection between two cities, with its weight corresponding to the number of cross-city trips. A directed edge is graphically depicted as an arrow indicating the direction of travel, from the starting city to the ending city.
- **Network diameter**: This is the shortest path between the two most distant nodes in a network. It indicates the number of steps required to travel from one side of the network to the other, providing insight into the network's overall size.
- **Graph density**: This measures the ratio of actual edges to all possible edges in a network. A complete network, where every node is connected to every other node, has a density of 1. In contrast, a null graph, in which all nodes are isolated and there are no edges, has a density of 0. A denser network facilitates cross-city travel more easily compared to a sparser network.
- **Average path length**: This is defined as the average number of steps along the shortest paths between all possible pairs of nodes. It reflects the efficiency of cross-city travel within the network.
- **Degree**: The degree of a node refers to the number of edges connected to it.
- **Weighted degree**: This is the sum of the weights of all edges connected to a node. Both **degree** and **weighted degree** are centrality measures, assigning numerical values to nodes to indicate their relative importance within the network.

In addition to the basic measurements mentioned above, this chapter further explores network characteristics by detecting **spatial communities**—sets of nodes that are more densely connected internally than they are with the rest of the network [22]. In this study, a **spatial community** refers to a group of cities where the connections between cities within the same community are stronger and more frequent than those with cities outside the community. Various algorithms have been developed to identify communities within complex networks [10, 14]. These algorithms aim to partition a network into groups of nodes with dense internal connections and relatively sparse connections between groups.

Spatial communities are identified using the Infomap algorithm [35]. According to existing literature, Infomap outperforms other commonly used algorithms when applied to weighted and directed networks [1, 41], which aligns with the characteristics of the cycling spatial structure derived from cross-city travel. Unlike modularity maximisation—one of the most widely adopted methods for community detection [22, 29]—the Infomap algorithm, introduced by Rosvall and Bergstrom [35, 34]), employs the Map Equation method to identify communities in directed and weighted graphs. The Map Equation is a flow-based, information-theoretic approach that "specifies the theoretical limit of how concisely we can describe the trajectory of a random walker on the network" [33]. By treating a random walker as a proxy for real flows, minimising the Map Equation across all possible partitions can uncover meaningful structural features of the network.

2.5 Results

2.5.1 Overall Bike-Sharing Usage

We adopt four main criteria to pre-process the dockless bike-sharing data. First, we exclude trips generated outside the geographic area of Zhejiang Province. Second, we exclude trips shorter than 50 m or longer than 50 km. Third, we exclude trips lasting less than 60 s or more than 10 h. And finally, we exclude trips with speeds slower than 8 m/min or faster than 1900 m/min. It is important to note that the selection of speed thresholds is based on the speed distribution of all original trips. Specifically, more than 99% of all original trips had a speed slower than 1,894 m/min (approximately 1,900 m/min), while only 1% of the original trips had a speed slower than 7.72 m/min (approximately 8 m/min). After pre-processing, we obtained a dataset containing 12.49 million trips, generated by 2.11 million users and 381.60 thousand bikes.

All prefecture-level cities have bike trips (Fig. 2.3). Hangzhou, as the provincial capital, has the largest number of trips (6.33 million), followed by Wenzhou (2.77 million) in the southeast and Ningbo (1.55 million) in the northeast. These three cities are also the top three in terms of population size and economic scale within Zhejiang Province. Users in Jiaxing and Shaoxing generate the smallest trip demand, with 38.66 thousand and 23.74 thousand trips, respectively. At a finer spatial scale (Fig. 2.3a), the urban cells, which have a 1-km width/length, in the central areas of prefecture- and county-level cities are more likely to experience higher trip volumes, such as those located in the city centres of Hangzhou and Ningbo.

At the county level, bike-sharing services are provided in 70 cities (i.e., 70 counties/districts in this chapter) (Fig. 2.3b). Districts, which typically cover the core urban areas of a prefecture-level city, are more economically vibrant and populous, and thus more likely to generate a higher trip volume. All of the top four county-level cities with the largest number of trips are districts located within the provincial capital, Hangzhou. These are Xihu (1.54 million), Shangcheng (1.36 million), Gongshu (1.22 million), and Qiantang (0.95 million). Conversely, some cities located in the western and southwestern areas have no or limited bike-sharing services.

As shown in Fig. 2.4a–c, the average trip distance is 1.52 km (with a standard deviation (SD) of 1.29 km), the average trip duration is 10.52 min (SD = 10.70 min), and the average travel speed is 176.62 m/min (SD = 126.31 m/min). These figures are generally consistent with existing studies [20, 45]. Figure 2.4d demonstrates a clear

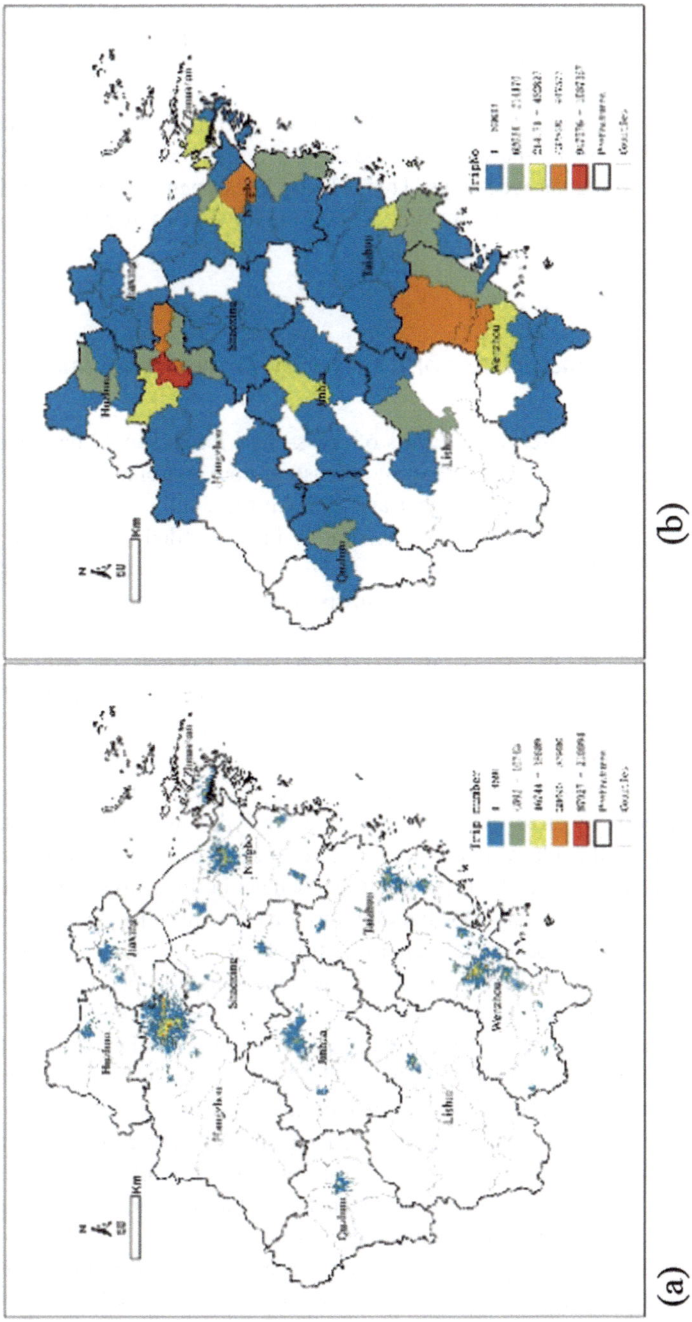

Fig. 2.3 Spatial patterns of dockless bike-sharing. Note: **A** represents the 1-km pattern whereas **B** represents the county-scale pattern

temporal regularity, with a higher number of trips observed during the morning and afternoon commuting peak times and on weekdays, compared to off-peak commuting times and weekends. It is worth noting that the lower trip numbers in the first week were likely due to rainy weather, particularly on September 14, 18, and 19, which has been frequently identified as an impact factor for micromobility [8]. On September 20 (Sunday), more users opted to ride bikes in the afternoon, possibly for leisure purposes. Despite it being the weekend, the days of September 26 and 27 followed a pattern similar to regular weekdays. This is likely because many people needed to work on these 2 days to make up for time lost during the seven-day national holiday starting on October 1.

2.5.2 *Basic Characteristics of Cross-City Travel Behaviour*

Table 2.1 presents the descriptive results of cross-city trips. It should be noted that Type II trips are not directly generated by the use of dockless bike-sharing, so it is not applicable to calculate bike number and average trip number per bike for Type II. Furthermore, since it is unlikely to accurately infer detailed information about how Type II cyclists travel between cities, we only roughly estimate the attributes of average trip distance, trip duration, and trip speed in order to understand the basic characteristics of Type II behaviour.

The results indicate that cross-city travel behaviour accounts for only a small proportion of all trips, particularly Type I trips (3401) at the prefecture level. The number of users and bikes associated with cross-city travel at the prefecture level is also considerably smaller than those for all trips. Cross-city travellers tend to have a lower average trip number (1.53 for Type I at the prefecture level), and bikes are used less frequently by cross-city travellers (1.73 for Type I at the prefecture level). This is likely due to the relatively infrequent occurrence of cross-city travel, given its larger spatial scale compared to inner-city travel behaviour. In terms of trip distance, duration, and speed, cross-city travellers tend to travel greater distances, for longer durations, and at higher speeds than inner-city travellers.

Cross-city behaviour is much more common at the county level compared to the prefecture level. For instance, at the county level, the number of Type I trips is 1.04 million, accounting for 8.3% of the total trip count, while the number of Type II trips is 1.27 million, representing 10.19% of the total. There are also more users and bikes travelling across county-level cities. However, the average trip distance, duration, and speed show an opposite trend, steadily decreasing from the prefecture-level to the county-level scale. In other words, riders travelling between prefecture-level cities travel longer distances (both in terms of distance and duration) and at faster speeds than those travelling solely within county-level cities.

Figure 2.5 shows the temporal distributions of cross-city trips at the prefecture and county levels. The Type I pattern at the county level (denoted as Type I (County) in Fig. 2.5a) generally follows a similar trend to Type I (Prefecture) across the days (as shown in Fig. 2.5a), although Type I (Prefecture) exhibits greater variability.

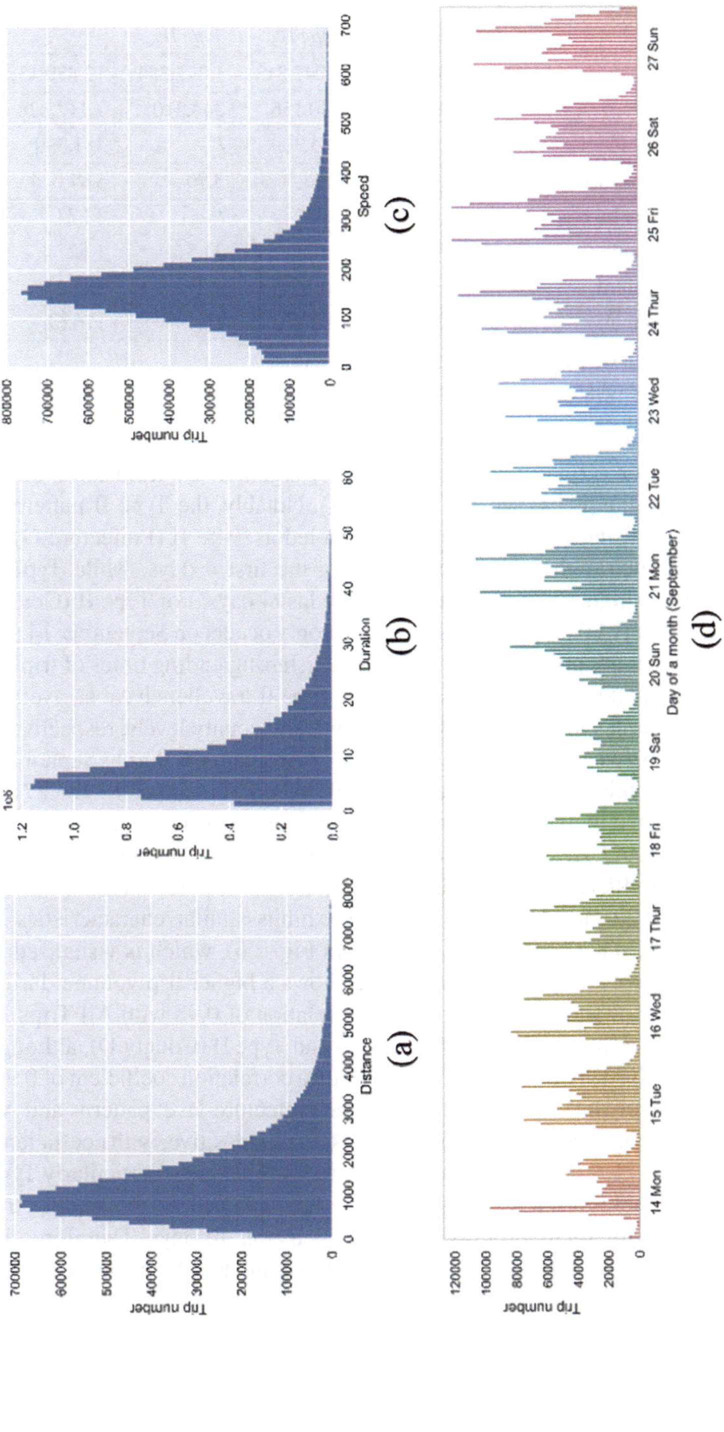

Fig. 2.4 Trip attribute histograms and temporal pattern of bike-sharing. Note: **a**, **b**, and **c** represent the histograms of trip distance, duration, and speed, respectively, and **d** shows a temporal trip pattern of bike-sharing

Table 2.1 Descriptive results of cross-city travel behaviour

	Prefecture level		County level		*All trips*
	Type I	*Type II*	*Type I*	*Type II*	
Trip number	3,401	27,859	1,042,313	1,272,384	12,489,138
User number	2,224	18,659	340,156	344,210	2,112,828
Bike number	1,968	/	165,309	/	381,599
Average trip number per user	1.53	1.49	3.06	3.70	5.91
Average trip number per bike	1.73	/	6.31	/	32.73
Average trip distance (km)	4.17	156.14*	2.67	21.51 *	1.52
Average trip duration (min)	26.48	3659.63*	16.56	1304.02*	10.52
Average trip speed (m/min)	198.93	42.67*	193.00	16.49*	176.62

/ shows it is not applicable to calculate a result in this position
* shows the numbers are roughly estimated

The patterns based on origin timestamps can differ significantly from those based on destination timestamps (shown in Fig. 2.5b, c). Notably, the Type II pattern at the prefecture-level using origin timestamps (denoted as Type II (Prefecture O) in Fig. 2.5b) has a much higher trip number during the first 4 days, while Type II (Prefecture D) has a higher trip number during the last 4 days. For Type II (County O) and Type II (County D), the temporal difference only occurs on September 14 and 27. These differences are primarily due to the use of starting/ending times of trips to delineate temporal patterns. Table 2.1 shows that Type II travellers have an average trip duration of 60.99 h and 21.73 h at the prefecture and county levels, respectively. The discrepancies between origin- and destination-based patterns during September 14 and 17 (or September 24 and 27) in Fig. 2.5b, or on September 14 (or 27) in Fig. 2.5b, are expected to diminish if data covering a longer period is included.

To better understand the association between temporal patterns, Fig. 2.6 shows their correlation coefficients. The coefficients between temporal patterns are all significant at the 0.001 level. Type I (County) exhibits similar characteristics to the pattern derived from all trips (i.e., All Trips in Fig. 2.6), which is visualised in Fig. 2.4d. In other words, Type I (County) also shows a higher trip volume during commuting peak times and weekends, with a correlation of 0.98 with All Trips. A similar pattern is observed for Type II (County O) and Type II (County D), although their association with All Trips is slightly lower, with a correlation coefficient of 0.95. At the prefecture level, the similarities between the prefecture-level patterns and All Trips decrease, although their general associations remain positive, with coefficients at least larger than 0.71. Additionally, the prefecture-level patterns, particularly Type I (Prefecture), are more unstable. This can partly be explained by the smaller trip volumes observed with a temporal resolution of 24 h over 14 days. Therefore, the patterns delineated by trip numbers exhibit greater fluctuations, which may be further amplified by rainy weather.

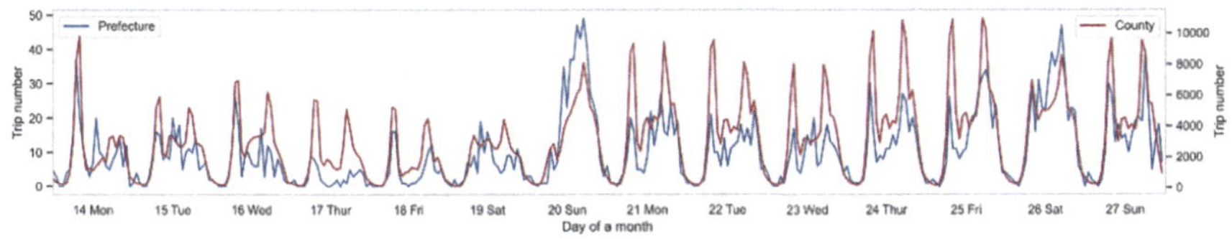

(a) *Type I (Prefecture)* and *Type I (County)*

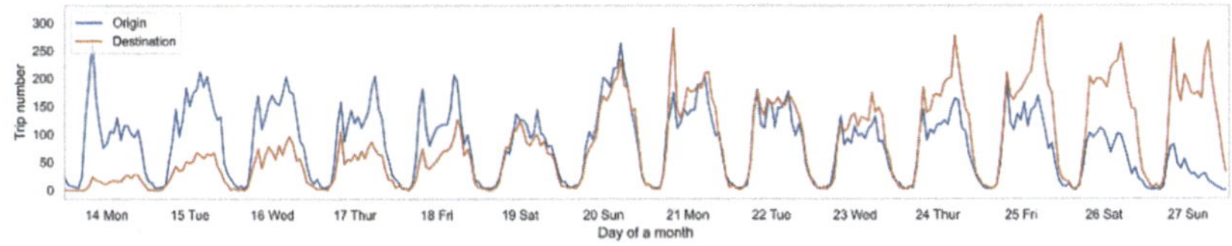

(b) *Type II (Prefecture)*

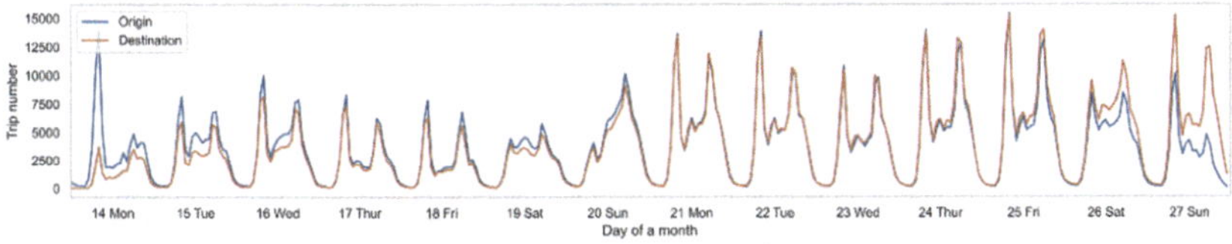

(c) *Type II (County)*

Fig. 2.5 Temporal distributions of cross-city travel behaviour

2.5.3 Associated Cycling Spatial Structures

This chapter quantitatively calculates several network attributes to describe the characteristics of the cycling spatial structures (Table 2.2). Among the 11 prefecture-level cities, 8 and 11 cities are connected in Type I (Prefecture) and Type II (Prefecture), respectively. Sixty-two out of 90 counties/districts are connected in county-level patterns. The Type II patterns and the county-level patterns have more edges. Type I (County) has the largest network diameter (6), indicating that there are 6 counties between the two most distant counties in its network. The average path length for Type I (County) is 1.92, also indicating that the distance between any two nodes in this pattern is, on average, the farthest. Type II (Prefecture) has a graph density of 0.96 (close to 1), demonstrating that it is almost a near-complete network. The networks are incomplete for the remaining three types of behaviour, but Type II (County) is more complete than the other two. Type II (County) has the smallest average distance (2) from a given starting node to all other nodes. The attributes mentioned above, such as network diameter, graph density, and average path length, generally indicate that the Type II patterns are more complex and complete, consisting of well-interconnected nodes (prefectures/counties), compared to the Type I patterns. The difference in complexity and completeness between the Type I and II patterns can be explained by the characteristics of cross-city travel behaviour. Type I travellers only travel across nearby cities, meaning connections are created between

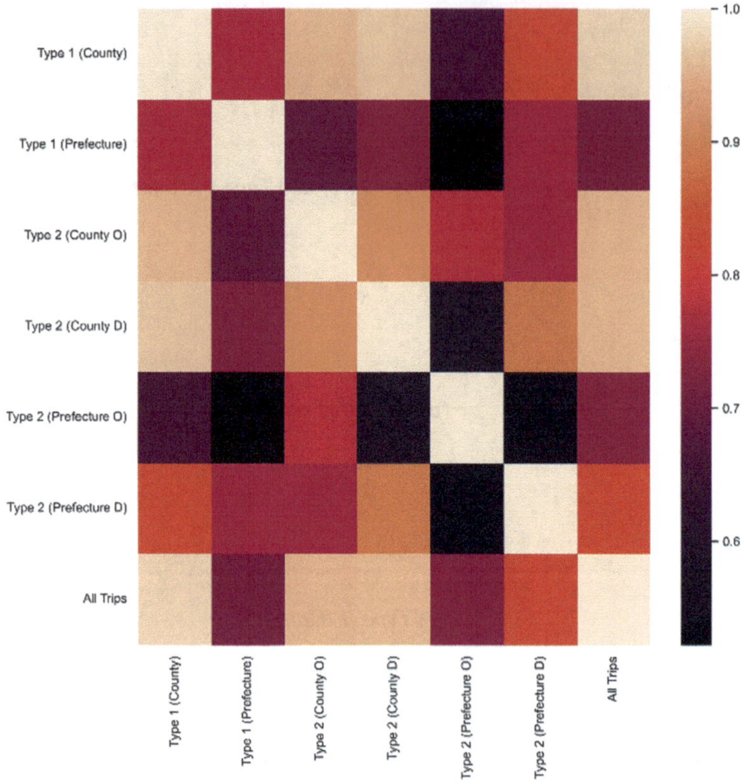

Fig. 2.6 The correlation heatmap across temporal patterns. Note: All correlation coefficients are significant at the 0.001 level

adjacent cities, while cities that are not adjacent have no opportunity to be directly connected via cycling. In contrast, Type II travellers are more likely to travel across cities with lower spatial proximity, thereby generating edges between more cities and resulting in a more complete and well-connected network. In terms of the degree measure, the average (weighted) degree increases as the number of nodes and edges increases. Type II (County) has the highest (weighted) degree. Each county in Type II (County) connects to 30 other counties on average, and its average weighted number of connections reaches 28,275.

Figure 2.7 shows the cycling spatial structures of cross-city travel behaviour using Gephi software and from a network perspective. In the Type I (Prefecture) pattern (Fig. 2.7a), only 8 of 11 prefecture-level cities are connected to at least one other city, and Hangzhou is strongly connected to its nearby city, Jiaxi (Fig. 2.7b). In the Type II (Prefecture) pattern (Fig. 2.7c), all prefecture-level cities are interconnected. Visually, Hangzhou is still the centre of the entire network, but it is more connected to other important cities such as Ningbo and Wenzhou. Regarding the county-level cities, we can observe several county clusters in Fig. 2.7d. The counties in the same

Table 2.2 Descriptive results of spatial networks from cross-city travel behaviour

Attributes	Prefecture-level		County-level	
	Type I	*Type II*	*Type I*	*Type II*
Number of nodes (connected nodes)	11(8)	11(11)	90(62)	90(62)
Number of edges	10	106	176	1,353
Network diameter	3	2	6	3
Graph density	0.09	0.96	0.02	0.17
Average path length	1.70	1.04	1.92	1.73
Average degree	1.82	19.27	3.91	30.07
Average weighted degree	618.36	5,065.27	23,162.51	28,275.20

cluster are usually within the same prefectural administrative boundary. In general, these clusters, with no connection to each other, are dispersed throughout the whole province. Despite there being 62 connected counties in both the Type I (County) and Type II (County) patterns, the counties are much more interconnected, with 1,353 edges in the latter, forming a more complex cycling spatial structure (Fig. 2.7e). In both Fig. 2.7d, e, we can easily recognise several counties with a large number of connections in Hangzhou, including the districts of Gongshu, Xihu, and Shangcheng.

This chapter uses the Infomap algorithm to detect spatial communities from the four networks. There are two detected communities in Fig. 2.8a. The larger one consists of 6 prefectures, characterised by a strong connection between Hangzhou and its nearby city, Jiaxing. The other community is located in the southeastern coastal area of Zhejiang Province, comprising only 2 prefectures (Taizhou and Wenzhou). In Fig. 2.8b, all prefectures are grouped into one community under the Type II behaviour across the prefecture cities. Compared to Fig. 2.8a, the prefecture-level cities are more interconnected, with a stronger connection between Hangzhou, Ningbo, and Wenzhou.

Figure 2.8c (or d) shows 10 detected communities. Most counties within a community are located within the administrative boundary of the same prefecture-level city. However, there are four exceptions: Communities 1, 2, 3, and 5. Community 1, located in the northern part, includes 14 counties/districts. The counties in Community 1 come from four prefecture-level cities: Hangzhou (10 counties), Shaoxing (2), Jiaxing (1), and Huzhou (1). Community 2 consists of 9 counties, including 8 from Ningbo and 1 from Shaoxing. Community 3 contains 15 counties, the largest number among all the communities. Its distribution is more spatially dispersed, including 8 counties from Wenzhou and 6 from Taizhou. Community 5 includes 3 counties, with two inside Quzhou's administrative boundary and one (Jiande) from Hangzhou.

Despite being visually more complex than Fig. 2.8c (or d), Fig. 2.8e (or f) contains a similar number of spatial communities (9). More communities show inconsistency between community-based and administrative boundaries. Community 1 has the largest number of counties (16). Compared to Community 1 in Fig. 2.8d, the changes are mainly caused by the counties from Huzhou, Jiaxing, and Shaoxing. Specifically,

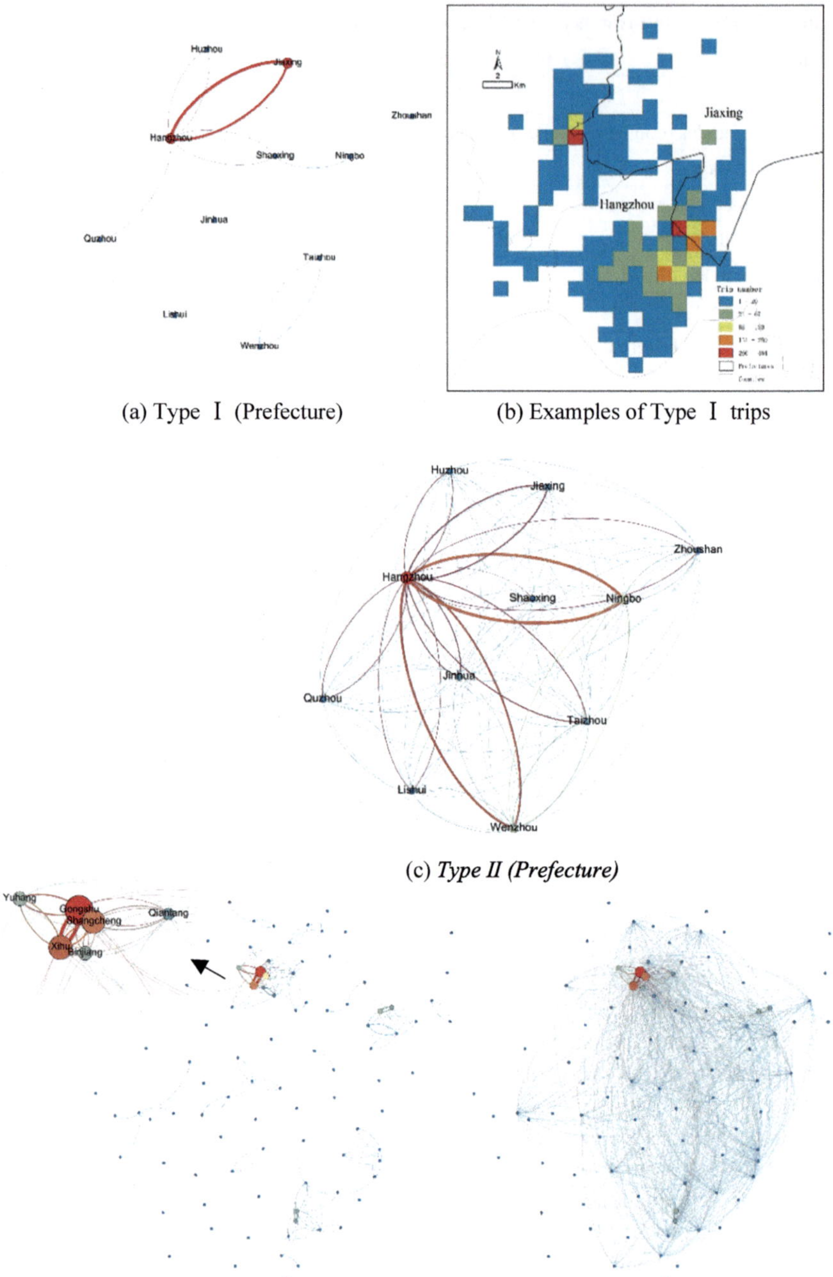

(a) Type Ⅰ (Prefecture)

(b) Examples of Type Ⅰ trips

(c) *Type II (Prefecture)*

(d) *Type I (County)*

(e) *Type II (County)*

Fig. 2.7 Cycling spatial structures of cross-city travel behaviour

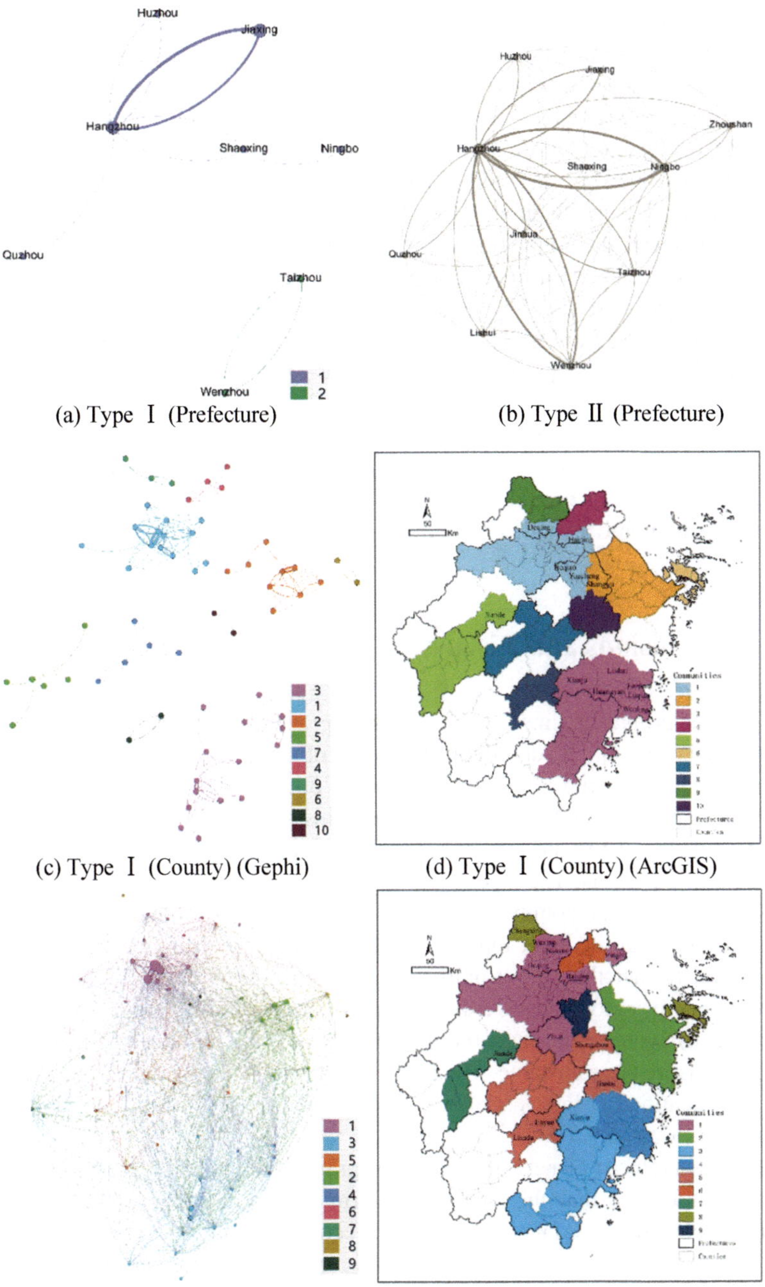

(a) Type Ⅰ (Prefecture)

(b) Type Ⅱ (Prefecture)

(c) Type Ⅰ (County) (Gephi)

(d) Type Ⅰ (County) (ArcGIS)

(e) Type Ⅱ (County) (Gephi)

(f) Type Ⅱ (County) (ArcGIS)

Fig. 2.8 Associated spatial communities

Nanxun and Wuxing counties from Huzhou belong to Community 1 in Fig. 2.8f but do not belong to Community 1 in Fig. 2.8d. Community 3 in Fig. 2.8d is divided into Communities 3 and 4 in Fig. 2.8f. Community 3 in Fig. 2.8f is the second largest community (11), with most of its counties located within Wenzhou's administrative boundary. Compared to those in Fig. 2.8d, Community 5 is newly formed. Its distribution is very diverse, consisting of 6 counties from Jinhua, 2 from Lishui, 1 from Shaoxing, and 1 from Taizhou. When comparing the Type I (County) and Type II (County) patterns, the communities in Type II (County) include counties that may not be geographically adjacent to each other, such as Community 5 and Community 8. Community 8 consists of 2 counties (Putuo and Dinghai) from Zhoushan and 1 county (Changxing) from Huzhou. Interestingly, Changxing is spatially very far from Putuo and Dinghai. The prefecture-level city of Zhoushan is an island city, making travel between Zhoushan and other cities more difficult. However, all three of these counties are famous for their tourism industry, attracting thousands and millions of visitors each year. As a result, some tourists may continuously visit these cities (as recorded in their cross-city trip data), forming connections between them and combining them into a single community.

2.6 Conclusions and Discussion

This chapter primarily aims to characterise the cycling spatial structures of cross-city travel behaviour. It identified two types of cross-city trips at the prefecture and county levels using a two-week dockless bike-sharing dataset from Zhejiang province, consisting of 12.49 million trips. This study then measured the cycling spatial structures using the Infomap algorithm and a set of indicators and methods from network analysis. The results showed that cross-city travel behaviour is more easily identified at the county level. Cross-city riders travel longer (in terms of trip distance and duration) than inner-city riders, and those crossing prefecture-level cities travel further than those only crossing county-level cities. From a temporal perspective, cross-city patterns exhibit typical characteristics of micromobility, such as a higher trip volume during commuting peak times and weekends. The prefecture-level patterns are more unstable, which may be attributed to the small number of cross-city trips identified at this spatial scale and the presence of occasional influencing factors (e.g., weather).

The measurement of network diameter, graph density, and average path length generally indicated that the Type II patterns are more complex and complete, consisting of well-interconnected nodes (prefectures/counties), compared to the Type I patterns. At the prefecture level, Hangzhou is positioned at the centre of the entire network of cities. It is strongly connected to its nearby city Jiaxi in Type I (Prefecture), and more connected to other important cities such as Ningbo and Wenzhou in Type II (Prefecture). At the county level, despite most counties within a spatial community being within the administrative boundary of the same prefecture-level city, some communities consist of counties that originally come from different prefecture-level cities and are even not geographically adjacent to each other. The communities in Type II (County) (Fig. 2.8f) consist of counties that may not be geographically adjacent to one another, whereas in Type I (County) (Fig. 2.8d), the counties within every community are adjacent. This is because, by definition, Type I travellers only travel across nearby cities, meaning only adjacent cities have the opportunity to be directly connected.

The presented study helps us better understand the characteristics of cross-city travel behaviour and cycling spatial structures. Cross-city travellers exhibit some different characteristics compared to inner-city travellers. To some extent, the cycling spatial structures can also reflect the relationships between cities and the role a city plays within a network. Various urban data sources, such as mobile phone data [23] and geolocated migration big data [43], have been used to reveal the characteristics of the systems of cities. This study demonstrates that bike-sharing data also have the potential to achieve this goal.

Our research also has some implications for policy-making and the practical operation of bike-sharing. Firstly, it can assist in designing subscription packages for bike-sharing service providers. It is interesting to explore the possibility of offering diverse price options depending on whether a subscription package grants the right to travel across multiple cities. An inner-city traveller would have no interest in enjoying the cross-city travel flexibility, so they could opt for an inner-city package, which would naturally be (slightly) cheaper than a cross-city package. On the other hand, the service provider could generate additional economic benefits from cross-city travellers by offering them the convenience of travelling across several cities in a province (e.g., between Hangzhou and Jiaxing in our case) or even across multiple cities nationwide.

Despite the results from our study showing that most bike-sharing users are inner-city travellers, we still observe that some users engaged in cross-city travel during the two weeks in September 2020. For instance, 0.11% and 16.10% of all users were cross-city travellers at the prefecture and county levels, respectively. It is expected that the number of cross-city travellers will increase after the COVID-19 pandemic, as the economy recovers and associated travel demand rises, including more cross-city business and leisure trips. In fact, in 2022, the company Hellobike introduced new options to distinguish cross-city packages from inner-city packages. For example, an annual membership limited to use within Hangzhou City is priced at 360 Yuan, while an annual membership that can be used nationwide is priced at 420 Yuan. The change in Hellobike's pricing packages clearly demonstrates the potential value of the cross-city travel behaviour discussed in this chapter.

The revealed cycling spatial structures may offer valuable guidance for re-designing operational areas. Traditionally, operational areas are divided based on administrative boundaries, with each city regarded as an independent operational area. However, our results indicate that administrative boundaries do not always align with actual cycling activity zones. This discrepancy has also been highlighted by some existing studies [5]. If there are many cross-city travellers between two cities, it would be beneficial to integrate these two cities (i.e., two operational areas) into a single operational area. Specifically, in our study, it is worth exploring the possibility of incorporating Jiaxing into Hangzhou as an integrated operational area. Other potential examples of re-designing operational areas include, but are not limited to, Deqing, Keqiao, Shangyu, Xianju, and Jiande.

There are some limitations to our study. Despite the original dataset being very large, containing 12.49 million trips, the number of cross-city trips identified is still limited. The data may not be entirely representative due to the influence of rainy weather. A dataset covering a longer period would be preferable to obtain a sufficient number of cross-city trips at different spatial scales and mitigate the effects of weather. Additionally, there are only 11 prefecture-level cities in Zhejiang Province, meaning that the cycling spatial structure at this scale is relatively simple. We could extend our discussion by considering more prefecture-level cities, such as by including nearby provinces in the Yangtze River Delta. Another issue with this study is that some districts or counties within the same prefecture-level city are very well-integrated functionally and economically, which means it is inappropriate to treat each district (or county) as a separate unit in our network analysis.

There are several future directions to improve our study. It would be interesting to quantitatively estimate the potential economic benefits of offering cross-city subscription packages, allowing us to determine exactly how to set the service price to meet the preferences of different users. In addition, we have now obtained e-bike-sharing data for the same period in Zhejiang Province. This data could be used to identify cross-city travel behaviour and cycling spatial structure derived from e-bike-sharing, and a comparison could be made with the bike-sharing data. Finally, it would be valuable to investigate the association between the built environment, socioeconomic factors, and cross-city travel behaviour, for example, by using geographically weighted regression models [19]. This investigation could provide useful insights for transport planning and policy, ultimately improving bike-sharing services.

References

1. Alzahrani T, Horadam KJ (2016) Community detection in bipartite networks: algorithms and case studies. In: Lü J, Yu X, Chen G, Yu W (eds) Complex systems and networks: dynamics, controls and applications. Springer, Berlin, pp 25–50
2. Bao J, He T, Ruan S et al (2017) Planning bike lanes based on sharing-bikes' trajectories. In: Proceedings of the 23rd ACM SIGKDD international conference on knowledge discovery and data mining, pp 1377–1386
3. Böcker L, Anderson E, Uteng TP, Throndsen T (2020) Bike sharing use in conjunction to public transport: exploring spatiotemporal, age and gender dimensions in Oslo, Norway. Transp Res Part Policy Pract 138:389–401. https://doi.org/10.1016/j.tra.2020.06.009
4. Caggiani L, Colovic A, Ottomanelli M (2020) An equality-based model for bike-sharing stations location in bicycle-public transport multimodal mobility. Transp Res Part Policy Pract 140:251–265
5. Chen W, Chen X, Cheng L et al (2022) Delineating borders of urban activity zones with free-floating bike sharing spatial interaction network. J Transp Geogr 104:103442
6. Chen Y, Zhang Y, Coffman D, Mi Z (2022) An environmental benefit analysis of bike sharing in New York City. Cities 121:103475

7. Elmashhara MG, Silva J, Sá E et al (2022) Factors influencing user behaviour in micromobility sharing systems: a systematic literature review and research directions. Travel Behav Soc 27:1–25. https://doi.org/10.1016/j.tbs.2021.10.001
8. Eren E, Uz VE (2020) A review on bike-sharing: the factors affecting bike-sharing demand. Sustain Cities Soc 54:101882
9. Filipe Teixeira J, Silva C, Moura e Sá F, (2022) The role of bike sharing during the coronavirus pandemic: an analysis of the mobility patterns and perceptions of Lisbon's GIRA users. Transp Res Part Policy Pract 159:17–34. https://doi.org/10.1016/j.tra.2022.03.018
10. Girvan M, Newman ME (2002) Community structure in social and biological networks. Proc Natl Acad Sci 99:7821–7826
11. Hellobike (2022) Hello 2021 sustainability report. https://www.163.com/dy/article/H96RPD320511BE1V.html. Accessed 4 Sept 2022
12. Heydari S, Konstantinoudis G, Behsoodi AW (2021) Effect of the COVID-19 pandemic on bike-sharing demand and hire time: evidence from Santander Cycles in London. PLoS ONE 16:e0260969
13. Hu Y, Zhang Y, Lamb D et al (2019) Examining and optimizing the BCycle bike-sharing system—a pilot study in Colorado, US. Appl Energy 247:1–12
14. Javed MA, Younis MS, Latif S et al (2018) Community detection in networks: a multidisciplinary review. J Netw Comput Appl 108:87–111. https://doi.org/10.1016/j.jnca.2018.02.011
15. Jiao J, Lee HK, Choi SJ (2022) Impacts of COVID-19 on bike-sharing usages in Seoul, South Korea. Cities 130:103849
16. Kaviti S, Venigalla MM, Lucas K (2019) Travel behavior and price preferences of bikesharing members and casual users: a Capital Bikeshare perspective. Travel Behav Soc 15:133–145. https://doi.org/10.1016/j.tbs.2019.02.004
17. Kong H, Jin ST, Sui DZ (2020) Deciphering the relationship between bikesharing and public transit: modal substitution, integration, and complementation. Transp Res Part Transp Environ 85:102392
18. Kou Z, Cai H (2019) Understanding bike sharing travel patterns: an analysis of trip data from eight cities. Phys Stat Mech Its Appl 515:785–797
19. Li A, Zhao P, Huang Y et al (2020) An empirical analysis of dockless bike-sharing utilization and its explanatory factors: case study from Shanghai, China. J Transp Geogr 88:102828
20. Lin D, Zhang Y, Zhu R, Meng L (2019) The analysis of catchment areas of metro stations using trajectory data generated by dockless shared bikes. Sustain Cities Soc 49:101598
21. Liu X, Pelechrinis K (2021) Excess demand prediction for bike sharing systems. PLoS ONE 16:e0252894
22. Liu Y, Zhang Y, Jin ST, Liu Y (2020) Spatial pattern of leisure activities among residents in Beijing, China: exploring the impacts of urban environment. Sustain Cities Soc 52:101806. https://doi.org/10.1016/j.scs.2019.101806
23. Louail T, Lenormand M, Cantu Ros OG et al (2014) From mobile phone data to the spatial structure of cities. Sci Rep 4:5276. https://doi.org/10.1038/srep05276
24. Lu Y, Prato CG, Corcoran J (2021) Disentangling the behavioural side of the first and last mile problem: the role of modality style and the built environment. J Transp Geogr 91:102936
25. Ma X, Ji Y, Yuan Y et al (2020) A comparison in travel patterns and determinants of user demand between docked and dockless bike-sharing systems using multi-sourced data. Transp Res Part Policy Pract 139:148–173
26. Manca F, Sivakumar A, Polak JW (2019) The effect of social influence and social interactions on the adoption of a new technology: the use of bike sharing in a student population. Transp Res Part C Emerg Technol 105:611–625. https://doi.org/10.1016/j.trc.2019.02.010
27. Martens K (2004) The bicycle as a feedering mode: experiences from three European countries. Transp Res Part Transp Environ 9:281–294
28. McKenzie G (2019) Spatiotemporal comparative analysis of scooter-share and bike-share usage patterns in Washington, DC. J Transp Geogr 78:19–28

29. Newman ME (2006) Modularity and community structure in networks. Proc Natl Acad Sci 103:8577–8582
30. O'Brien O, Cheshire J, Batty M (2014) Mining bicycle sharing data for generating insights into sustainable transport systems. J Transp Geogr 34:262–273. https://doi.org/10.1016/j.jtrangeo.2013.06.007
31. O'Sullivan F (2022) The radical roots of Bikesharing. In: Pocket. https://getpocket.com/explore/item/the-radical-roots-of-bikesharing. Accessed 22 Nov 2022
32. Radzimski A, Dzięcielski M (2021) Exploring the relationship between bike-sharing and public transport in Poznań, Poland. Transp Res Part Policy Pract 145:189–202
33. Rosvall M, Axelsson D, Bergstrom CT (2009) The map equation. Eur Phys J Spec Top 178:13–23
34. Rosvall M, Bergstrom CT (2007) An information-theoretic framework for resolving community structure in complex networks. Proc Natl Acad Sci 104:7327–7331
35. Rosvall M, Bergstrom CT (2008) Maps of random walks on complex networks reveal community structure. Proc Natl Acad Sci 105:1118–1123
36. Sadeghinasr B, Akhavan A, Furth PG et al (2021) Mining dockless bikeshare data for insights into cyclist behavior and preferences: evidence from the Boston region. Transp Res Part Transp Environ 100:103044
37. Shen Y, Zhang X, Zhao J (2018) Understanding the usage of dockless bike sharing in Singapore. Int J Sustain Transp 12:686–700
38. Sohrabi S, Ermagun A (2021) Dynamic bike sharing traffic prediction using spatiotemporal pattern detection. Transp Res Part Transp Environ 90:102647
39. Su D, Wang Y, Yang N, Wang X (2020) Promoting considerate parking behavior in dockless bike-sharing: an experimental study. Transp Res Part Policy Pract 140:153–165. https://doi.org/10.1016/j.tra.2020.08.006
40. Wang K, Akar G, Chen Y-J (2018) Bike sharing differences among Millennials, Gen Xers, and Baby Boomers: lessons learnt from New York City's bike share. Transp Res Part Policy Pract 116:1–14. https://doi.org/10.1016/j.tra.2018.06.001
41. Wu C, Smith D, Wang M (2021) Simulating the urban spatial structure with spatial interaction: a case study of urban polycentricity under different scenarios. Comput Environ Urban Syst 89:101677
42. Yang Y, Heppenstall A, Turner A, Comber A (2019) A spatiotemporal and graph-based analysis of dockless bike sharing patterns to understand urban flows over the last mile. Comput Environ Urban Syst 77:101361. https://doi.org/10.1016/j.compenvurbsys.2019.101361
43. Yang Z, Hua Y, Cao Y et al (2022) Network patterns of Zhongyuan urban agglomeration in China based on Baidu migration data. ISPRS Int J Geo-Inf 11:62. https://doi.org/10.3390/ijgi11010062
44. Zaltz Austwick M, O'Brien O, Strano E, Viana M (2013) The structure of spatial networks and communities in bicycle sharing systems. PLoS ONE 8:e74685
45. Zhang Y, Lin D, Liu XC (2019) Biking islands in cities: an analysis combining bike trajectory and percolation theory. J Transp Geogr 80:102497
46. Zhang Y, Mi Z (2018) Environmental benefits of bike sharing: a big data-based analysis. Appl Energy 220:296–301
47. Zhang H, Zhuge C, Jia J et al (2021) Green travel mobility of dockless bike-sharing based on trip data in big cities: a spatial network analysis. J Clean Prod 313:127930
48. Zhejiang China (2022) About—overview—economy. http://www.ezhejiang.gov.cn/2022-01/20/c_392874.htm. Accessed 4 Sept 2022

Chapter 3
Infer Trip Purposes of Shared Micromobility Using a LDA-Based Analytical Framework

3.1 Introduction

Intensive urbanisation is occurring worldwide, leading to various urban challenges such as traffic congestion, air pollution, and public health crises [17, 41]. One approach to mitigate these urban issues is to promote a shift towards low-carbon mobility within urban transport systems. Against this backdrop, a relatively new concept has emerged in the transport field: shared micromobility services (SMSs). SMSs typically refer to shared, lightweight, low-speed, and eco-friendly transport modes, exemplified by bike-sharing systems. These systems have demonstrated multiple benefits, including addressing first- and last-mile connectivity, reducing carbon emissions by discouraging private car use, and promoting physical activity among urban residents, thereby contributing to the alleviation of urban problems [34, 35, 49]. The first bike-sharing system, the "White Bicycle" programme, was introduced in Amsterdam, the Netherlands, in 1965. Since then, bike-sharing systems have experienced rapid global expansion, particularly over the past decade. More recently, the emergence of electric bikes (e-bikes) and electric scooters (e-scooters) sharing systems has further diversified SMSs, making them essential components of urban transport networks and facilitating residents' daily socio-economic activities.

The rapid growth of SMSs has raised a series of issues concerning urban planning regulations, their influence on the broader urban transport system, and their socioeconomic and environmental impacts [1, 32]. An essential prerequisite for addressing these issues is understanding the mobility patterns of SMSs, as their impacts are largely reflected in how they reshape the mobility dynamics of cities. Urban mobility patterns essentially stem from the complex interactions between urban dwellers and intricate traffic systems, shaped by various trip purposes (e.g., commuting, cultural, and recreational activities). Therefore, investigating the trip purposes associated with micromobility can provide a more comprehensive understanding of urban mobility. This understanding also contributes to resolving SMS-related planning and regulation challenges, such as predicting bike-sharing demand under different scenarios,

Y. Zhang, *Understanding Shared Micromobility in China Using Massive User-Generated Trip Data*, Urban Sustainability,
https://doi.org/10.1007/978-981-96-8492-2_3

thereby supporting the development of sustainable urban micromobility transport systems.

Although several studies have focused on mobility patterns, particularly the spatiotemporal characteristics of SMSs [8], research on the trip purposes of SMSs remains limited [6, 37, 43]. Recently, the advent of dockless SMSs has provided an unprecedented data source—passively collected large-scale trajectory data—enabling a more refined understanding of mobility patterns, especially the trip purposes of SMSs, at a finer spatiotemporal granularity than ever before. The dockless nature of these services offers users greater flexibility, allowing them to travel as close as possible to their destinations. Such door-to-door services have facilitated a more accurate identification of trip purposes, which can be revealed through large-scale trip data mining. By integrating travel data—particularly the locations and timestamps of trip origins and destinations—with auxiliary data such as Points of Interest (POIs), land use, and land cover maps, the trip purposes of SMSs can be inferred using various methods, including the Bayesian rule-based method, the Markov chain model, and random forests [4, 18].

Undeniably, the existing research on the trip purpose inference for SMSs has certain limitations. Firstly, although numerous studies have investigated the imputation of trip purposes for SMSs, their findings often lack consistency. Furthermore, current research predominantly focuses on a single mode of SMSs at a time, with little attention given to comparative analyses of different SMS modes. Such comparisons are crucial for comprehensively understanding the usage patterns of SMSs and formulating targeted planning and operational strategies for various types of micromobility. Consequently, in light of these research gaps, it is essential to thoroughly examine trip purposes of SMSs and conduct comparative analyses between different SMS types using emerging data mining methods.

This study aims to infer and compare trip purposes for two types of SMSs: shared bikes and shared e-bikes. Specifically, a Latent Dirichlet Allocation (LDA)-based analytical framework will be employed. Trip purposes will be inferred by combining structured travel patterns extracted through the LDA model with ancillary data, such as POIs. The main contributions of this study are threefold. Firstly, an LDA-based analytical framework is proposed to uncover latent travel patterns and identify trip purposes for dockless shared bikes and e-bikes in Ningbo. Secondly, this study is the first to systematically compare trip purposes between different SMS modes—shared bikes and e-bikes—at an aggregated level. Thirdly, tailored planning and operational recommendations are provided for various trip purposes and scenarios for shared bikes and e-bikes, with the goal of supporting sustainable development of dockless micromobility systems in the city.

3.2 Literature Review

The rapid development of SMSs worldwide has garnered increasing academic attention. The majority of existing studies on SMSs have focused on four main areas: the description and profiling of spatio-temporal characteristics [13, 23]; the exploration of factors influencing SMS usage [27, 39]; planning and operational challenges, including fleet scheduling and geofencing area planning [12, 48] and the relationships between different shared micromobility modes—such as substitution, competition, and complementarity—as well as their interactions with other transport modes like car-hailing and public transport [7, 25]. Moreover, there remains a scarcity of research addressing critical issues such as inequities in SMS accessibility [10], carbon emissions and environmental impacts [29], and user safety.

3.2.1 Controversy and Limitations of Current Studies

Technically, research on the inference of trip purposes for SMSs falls within the broader category of understanding SMS usage, a topic that has recently received increasing scholarly attention. However, there is little consensus in the current literature. For instance, Bao et al. [6] investigated the trip purposes of docked shared bikes by analysing a smart card dataset from Citi Bike in New York City. Their findings indicated that dining out is the most common reason for bike-sharing trips, followed by transport and shopping. Similar results were reported by Xing et al. [43], who used K-means clustering to uncover dockless bike-sharing patterns and trip purposes in Shanghai. However, their approach had limitations, as clustering analysis cannot precisely identify the true purpose when a cyclist's destination is a POI encompassing multiple potential activities (e.g., integrated locations or commercial complexes). Li, Huang, et al. [26] partially addressed this limitation by employing a Dirichlet multinomial regression topic model to infer the destination activities of dockless bike-sharing trips in Shanghai, leading to more accurate trip purpose estimations. Nevertheless, their results showed slight discrepancies from those of Xing et al. [43], revealing that commuting was the most prevalent purpose for dockless shared bike trips, followed by dining out.

The inference of trip purposes for shared e-bikes and e-scooters has also attracted scholarly attention. Guidon et al. [16] analysed eight months of transaction data from a dockless e-bike sharing system in Zürich, Switzerland, revealing that commuting is the primary trip purpose. In contrast, He et al. [20] examined historical travel data from the Summit Bike Share system in Park City, Utah, and found that e-bike sharing is primarily used for recreation by casual users. Furthermore, McKenzie [33] demonstrated that dockless e-scooter trips in Washington D.C. are closely associated with recreational activities. Meanwhile, Zhu et al. [51] observed that in Singapore, e-scooters are predominantly used by college students commuting between metro stations and their campuses. Overall, there is little consensus in the current literature regarding the trip purposes of SMSs. These contradictions may stem from the markedly different urban geographical contexts and the rapid evolution of SMSs.

Aside from the discrepancies in their findings, most existing studies have focused solely on investigating the trip purposes of users relying on a single type of micromobility, with limited research comparing trip purposes across various micromobility modes. Such comparative analyses are crucial for the refined and targeted governance of different SMSs under diverse trip-purpose scenarios. Specifically, current comparative studies on different SMSs primarily address travel patterns and characteristics [3], user profiles [8], level of services [22], and safety risks [28], while comparisons of trip purposes remain largely overlooked. More recently, Reck et al. [36] conducted a comparative analysis of four distinct shared micromobility services in Zurich, Switzerland, examining micromobility mode choices across various trip purposes and destinations. Their findings revealed that mode choice is predominantly influenced by trip distance and time of day, with docked modes being favoured for commuting trips. Overall, their research offers a valuable starting point for further exploration in this area.

3.2.2 The Methods Used to Infer Trip Purposes

Considerable effort has been devoted to developing methods for inferring trip purposes of SMSs. Traditionally, scholars have primarily relied on questionnaire surveys to study trip purposes. For example, Shaheen et al. [40] investigated the use of bike-sharing systems in North America through an online survey, finding that most people use shared bikes for commuting to work or study. Similarly, Weis et al. [42] analysed survey data from the Mobility and Transport Microcensus (MTMC) in Switzerland, concluding that there is a correlation between routine choices, urban mobility modes, and trip purposes. However, critics have highlighted that such conventional surveys are costly, time-consuming, and may involve a degree of subjectivity, resulting in relatively high costs but low accuracy [24].

With the rapid development of the big data technology, scholars have gained the opportunity to reveal trip purposes in a cost-efficient manner by developing various data mining methods. A series of rule-based methods have been proposed, grounded in the fundamental concept that trip purposes are constrained by a range of soft factors such as space, time, and the sequence of activities. For instance, Zou et al. [52] identified trip purposes by examining the internal spatiotemporal relationships within multi-day smart card transaction data from Beijing's subway. Similarly, Zhang et al. [46] distinguished trips, trip chains, and transition activities using five months of bike-sharing trip data from Zhongshan, China, revealing that most cyclists use bikes for commuting. The inference of trip purposes not only relies on spatial and temporal constraints but can also be enhanced by auxiliary data sources such as POIs, land use and land cover maps, and mobile signalling data [38, 45]. For example, Gong et al. [15] developed a Bayesian rule-based model combined with a Monte

Carlo simulation to estimate the probability of visiting different POIs for taxi trips. Building upon Gong's method, Ross-Perez et al. [37] uncovered trip purposes of dockless bike-sharing trips in Manchester, finding that trips to residential areas are predominant.

Recently, to adapt to the ever-growing complexity of rules used in the study of the trip purpose inference, scholars have sought to apply sophisticated machine learning methods capable of automatically identifying potential features in large-scale data. These methods can be broadly categorised into supervised and unsupervised approaches. The former generally requires training with the assistance of auxiliary datasets (e.g., travel surveys). For instance, Ermagun et al. [14] developed a trip-purpose prediction model based on nested logit and random forest algorithms, incorporating trip information with Google Places API to deduce trip purposes. Similarly, Alsger et al. [4] devised a model to investigate passengers' trip purposes using smart card fare data from Brisbane, Queensland. Their model was calibrated and validated using multivariate data, including Household Travel Surveys and land use data. Compared to supervised machine learning models—which necessitate auxiliary datasets with predefined recognition labels—unsupervised machine learning methods are regarded as more cost-efficient and are thus increasingly employed in the trip purpose inference.

Among unsupervised machine learning methods, topic modelling—particularly Latent Dirichlet Allocation (LDA)—has gained popularity as it effectively extracts latent spatiotemporal dynamics of activities without relying on predefined keywords or grammatical logic. This offers unique advantages over earlier methods. For example, previous techniques such as Monte Carlo simulations and hidden Markov chain models largely depend on probabilities to infer trip purposes. However, these methods may lead to convergence in inferred trip purposes for closely situated destinations, as travellers are more likely to select locations with the highest probability scores.

Furthermore, studies employing K-means clustering have computed trip purposes by examining origins and destinations separately, failing to provide a comprehensive analysis of structured travel patterns integrating origin–destination pairs. In contrast, LDA-based methods synthesise origin–destination pairs and extract structured travel patterns from extensive shared micromobility records. For instance, Bao et al. [6] applied the LDA algorithm to analyse structured travel patterns and uncover the trip purposes of Citi Bike users in New York. These studies, to some extent, bridged the methodological gap and highlighted the potential and value of the LDA-based topic modelling in the trip purpose inference for SMSs.

Drawing on existing studies, it can be concluded that, despite substantial efforts to understand SMSs, research concerning the trip purpose inference and the comparison of trip purposes across different micromobility services remains limited. Latent Dirichlet Allocation, a topic modelling technique, has proven effective in inferring trip purposes efficiently. Building on this, the presented study seeks to apply an LDA-based analytical framework to infer trip purposes for two types of dockless SMSs: shared bikes and e-bikes.

3.3 Methodology

3.3.1 Study Area

Ningbo is a major city located in the northeast of Zhejiang Province, China. This study focuses on the central area of Ningbo, which includes five administrative districts: Zhenhai, Beilun, Haishu, Jiangbei, and Yinzhou (see Fig. 3.1). According to the Ningbo Statistical Yearbook, these five districts had a combined population of nearly one million and a total gross domestic product of 269 million yuan in 2019, making them the most densely populated and urbanised areas in Ningbo. Since 2016, this city has experienced a surge in the number of dockless shared vehicles, significantly impacting urban transport [19]. The close relationship between Ningbo and SMSs provides us with a unique opportunity to gain valuable insights into the role of micromobility within the city.

3.3.2 Data

This study utilises two primary datasets: a dockless bike-sharing travel dataset and a POI dataset. The dockless bike-sharing dataset, which includes both shared bikes and e-bikes, is provided by Hellobike, one of the largest internet technology companies operating bike-sharing systems in China. The dataset comprises 616,638 trips made in Ningbo, spanning from 14 to 20th September 2020. Each trip record includes the order ID, bike ID, anonymous user ID, bike type, the latitudes and longitudes of the pick-up and drop-off points, and the corresponding times. Trips with durations of less than 1 min or greater than 2 h are excluded, following similar criteria used in previous studies [47]. After this preprocessing, a total of 598,431 trips are retained as our study sample. As shown in Fig. 3.1, bike-sharing trips are predominantly concentrated around the urban centre rather than being evenly spread across the city. To better illustrate the travel characteristics of the two bike types, the subsequent analyses will focus on their respective concentration areas.

The POI data is sourced using application programming interfaces (APIs) provided by AutoNavi (also known as Gaode Map), a leading navigation company in China. This dataset includes 391,571 POIs in the central area of Ningbo, with related information such as names, geographical coordinates, and official function categories (e.g., residential buildings). However, the official POI classification is proved somewhat cumbersome for the trip purpose inference. Generally, travel purposes encompass recurring activities like commuting and schooling, as well as casual activities such as shopping, dining, and seeking services. By combining common travel purposes identified in existing literature with the official POI classification provided by AutoNavi, the POIs are reclassified into 11 types, including various common urban functions such as residential, shopping, dining, and more (see Fig. 3.2).

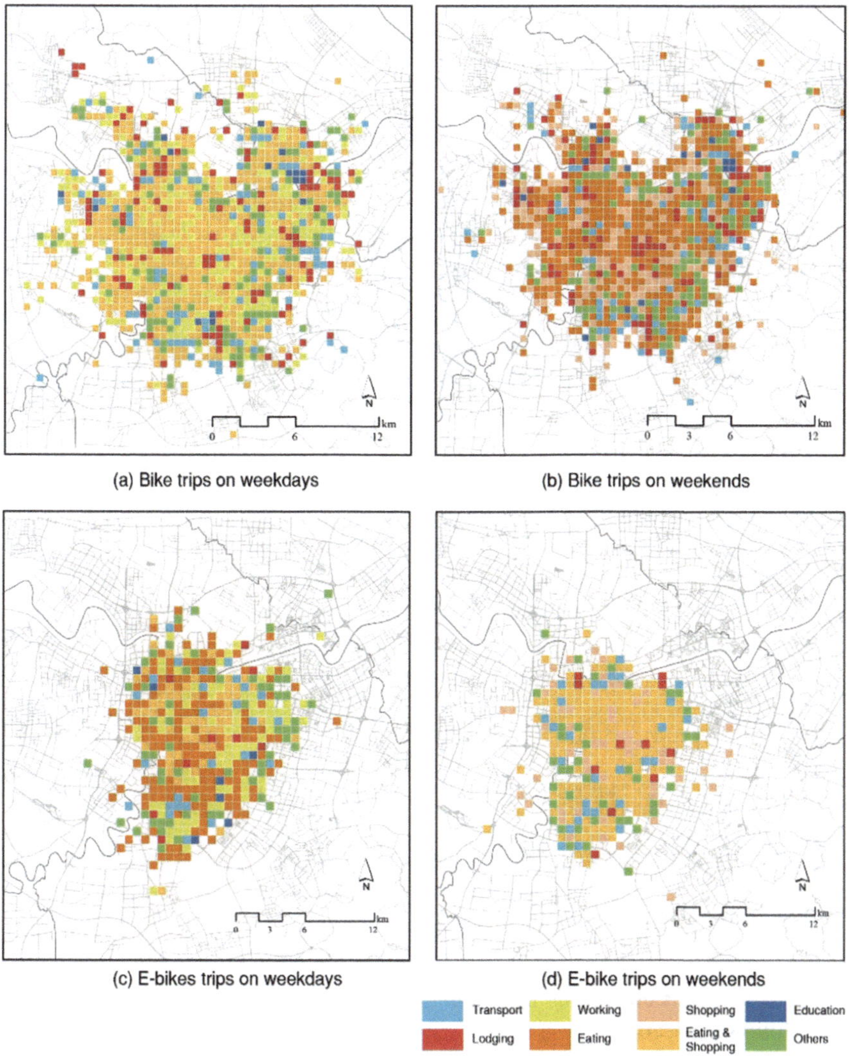

Fig. 3.1 The city of Ningbo

3.3.3 *An LDA-Based Analytical Framework to Infer Trip Purposes*

Building on existing studies, this chapter proposes an analytical framework to infer the trip purposes of cyclists (see Fig. 3.3). The framework comprises two main steps: trip annotation and the LDA model. The first step involves three sub-steps: defining destination areas, calculating POI feature vectors, and categorising bike-sharing trips

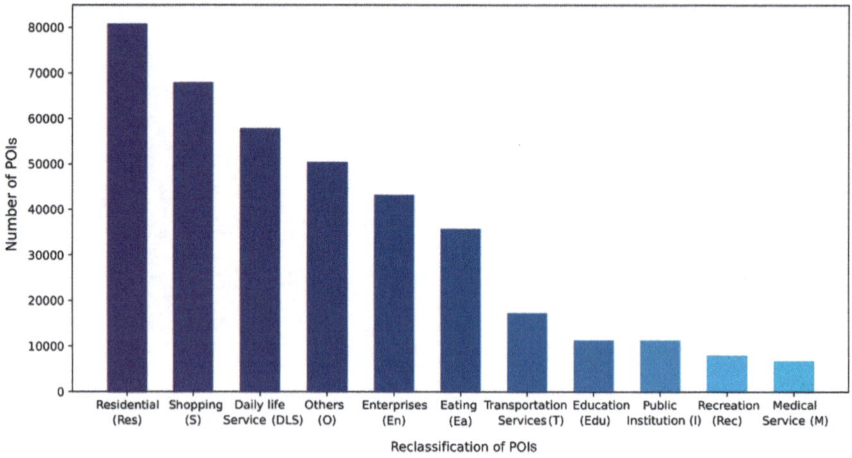

Fig. 3.2 Reclassification of POIs

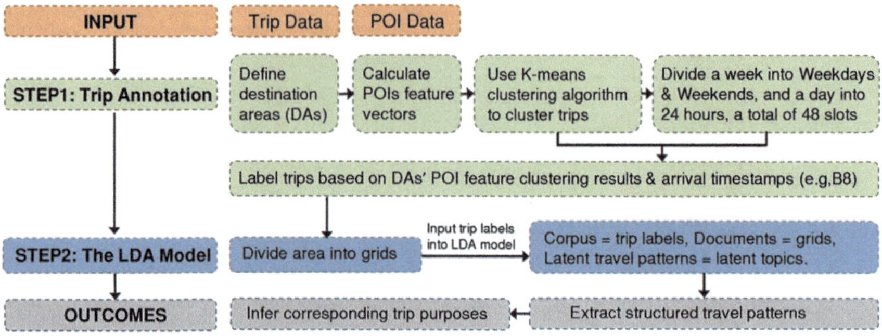

Fig. 3.3 Analytical framework

using the K-means clustering algorithm. The framework will be discussed in detail in the following sections.

3.3.3.1 Define Destination Areas

This study assumes that dockless bike-sharing users prefer to park their bikes as close to their destinations as possible. Identifying the connection between drop-off points and destinations is crucial for inferring trip purposes. However, the geo-coordinates of drop-off points cannot be directly used as the cyclists' destinations for two reasons. First, there is a location deviation between the potential destination and the drop-off point, as users typically park their dockless bikes in designated parking areas (such as geofenced zones) before walking to their final destination. Second, due to technical

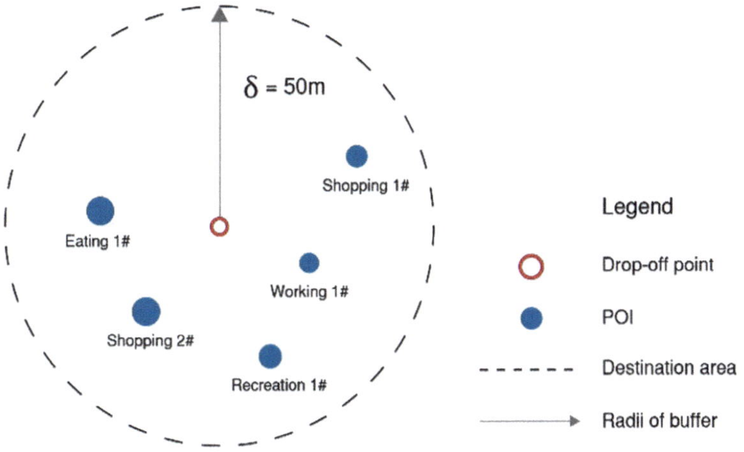

Fig. 3.4 The destination area

limitations, GPS errors are common; the geo-coordinates of origins and destinations recorded by the equipment are often slightly offset from their actual locations.

Taking the above considerations into account, this study defines a destination area for each trip using a buffer to accurately establish the connection between the drop-off point and the intended destination (see Fig. 3.4). Different buffer radii (δ) ranging from 10 to 100 m, with 10 m increments, are examined. The percentage of buffers covering at least one POI is calculated using the method proposed by Gong et al. [15]. We find that the percentage remained stable at 96% when δ reached 50 m. As a result, $\delta = 50$ m is chosen as the distance threshold, and 574,603 trips are retained as our final sample.

3.3.3.2 Calculate Feature Vectors for POIs' Distribution

After defining the destination area, the distribution characteristics of surrounding POIs at the trip's drop-off point can be calculated. Each trip is represented by a POI feature vector. For example, the proportion of the 11 types of POIs within the 50 m buffer around the drop-off point can be expressed as $T_1 = (c_1, c_2 \cdots\cdots c_{11})$, where $c_1, c_2 \cdots\cdots c_{11}$ represent the ratios of the number of each of the 11 types of POIs within the trip's destination area.

3.3.3.3 Categorise Shared Micromobility Trips

The POI feature vectors are input into the K-means clustering model to categorise trips. K-means clustering is an efficient unsupervised machine learning algorithm commonly used to divide a dataset into a specified number of K clusters by minimising intra-cluster distances and maximising inter-cluster distances [31]. Additionally, the algorithm provides clustering centres, each representing a central point that most accurately characterises the cluster and reflects its distinctive features. In this study, K-means clustering is applied to categorise bike-sharing trips based on their POI feature vectors. Specifically, trips with similar POI distribution feature vectors at their destinations are grouped into the same cluster.

The key to obtaining an ideal clustering result is determining an appropriate K value, which controls the number of clusters. In this study, the silhouette coefficient is used to identify the optimal K value. The silhouette coefficient is calculated as follows:

$$s_{(i)} = \frac{b_i - a_i}{\max(b_i, a_i)} \tag{3.1}$$

where, a_i is the mean distance between point i and all other points in the same cluster. b_i means the mean distance between point i and all other points in the next nearest cluster.

3.3.3.4 Annotate Trips with Labels

In this sub-section, this study annotates the trips with trip labels, which are discrete variables that can be input into the LDA model. First, trips are annotated based on the clustering results obtained in Sect. 3.3.3. The underlying logic is that trips with the same purpose are likely to have destinations with similar properties. Second, the arrival timestamps are taken into account, as trip purposes may vary depending on the time of arrival. For instance, those visiting a commercial area around noon are more likely to be there for food rather than work. For example, if the K-means clustering method grouped trips with different destinations into three categories, these are labelled as A, B, and C. If a cyclist arrives at a destination with a POI distribution characteristic of type B at 8:00 am on Monday, this trip would be annotated as B8.

3.3.3.5 Extract Travel Patterns via LDA

The LDA model is a topic modelling technique widely used in natural language processing. It has been frequently applied to extract latent topics from large volumes of documents, based on the concept that documents in a corpus are a mixture of topics, with each word supporting specific topics within the documents [9]. The LDA model is trained to count term frequencies, which enables the extraction of latent topics from a document. Ultimately, it produces two probability distribution matrices: one for documents-to-topics and another for topics-to-words.

In this chapter, the LDA model draws an analogy between extracting latent topics from documents and uncovering hidden trip patterns from trip records. The study area is divided into 500 m by 500 m grids, following the approach outlined by Chang

Table 3.1 The analogy between travel pattern identification and original LDA model

Case study	Original LDA model
All trip labels	Corpus
Grids	Documents
Latent travel patterns	Latent topics of a document
Trip labels in a grid	Words in a document

et al. [11]. Naturally, each grid contains several trips, along with their corresponding trip labels. In this analogy, grids represent documents, bike-sharing travel patterns represent topics, and trip labels represent words. The details of this analogy are presented in Table 3.1.

The general process of the LDA model is:

For each grid $G_d \epsilon \{ G_1, G_2, \ldots, G_D\}$, we draw a travel pattern distribution parameter vector $\theta_d \sim Dirichlet(\alpha)$.

For each latent travel pattern $k \epsilon \{ k_1, k_2, \ldots, k_N\}$, we draw a trip label distribution parameter vector $\varnothing_k \sim Dirichlet(\eta)$.

For each travel label $i \epsilon I$ in Grid G_d:

Draw a travel pattern $z_{d,i}$~ Multinomial (θ_d);

Draw a trip label $\omega_{d,i}$~ Multinomial ($\varnothing_{Z_{d,i}}$).

Where, D is the number of grids, N is the number of latent bike-sharing travel patterns, θ and $\varnothing$ are mixture weights matrix for each travel pattern k in each grid G_d, each trip label i in each travel pattern k, with the hyper-parameters α and η, respectively.

In this model, three parameters need to be pre-specified: K (the number of topics), α and η (the hyper-parameters of two Dirichlet distribution matrices). The Variational Expectation–Maximisation (VEM) algorithm, which can automatically initialise the hyperparameters of the model based on sufficient statistics [9], is adopted to obtain optimal α and η. In addition, the number of topics, K, for this study, that is the number of latent travel patterns, is a significant input which greatly impacts the LDA's performance. This study computes the perplexity to determine the optimal K value, calculated as follows:

$$Perplexity(D) = \exp\{-\frac{\Sigma_{d=1}^{M} \log pw(d)}{\Sigma_{d=1}^{M} N(d)}\} \tag{3.2}$$

where, D represents the corpus of trip labels, M is the number of grids, $N(d)$ is the number of trip labels in each grid, $pw(d)$ means the produce probability of trip label $w(d)$ in grids.

3.4 Results

3.4.1 Exploratory Analysis

Figure 3.5a and b illustrate the temporal distribution of bike-sharing trips on weekdays and weekends, respectively. Both shared bike and e-bike trips exhibit clear peaks in the morning and evening, with a noticeable drop in volume at 3 am. Regarding the proportion of shared bike and e-bike usage at the hourly level (Fig. 3.5c), shared bikes dominate most of the time, although e-bikes are more prevalent between 3 and 5 am, highlighting their importance for night-time travel. Figure 3.5d shows that the majority of shared bike trips are shorter than 2 km, while most e-bike trips range between 1.5 and 3 km. Figure 3.5e indicates that most shared bike and e-bike trips are completed within 30 min, accounting for 96% and 94% of all bike and e-bike trips, respectively. However, the histogram for e-bikes is more left-skewed than that for shared bikes, suggesting that the average trip duration and distance for e-bike trips are longer than those for shared bike trips. Figure 3.5f presents the speed distribution of shared bike and e-bike trips. Overall, the speed difference between the two types is not significant. The average speed of a shared bike is 2.7 m/s, while that of an e-bike is 2.91 m/s. An interesting observation from the histogram is that the speed distribution for shared bikes is more concentrated, whereas that for e-bikes shows greater variation. This may be due to the fact that riders of regular bikes tend to maintain a consistent pace for energy efficiency, while e-bike users have more flexibility to adjust their speed by controlling the electric power system.

3.4.2 Shared Micromobility Trips Categorisations

This chapter applies the K-means clustering algorithm to categorise bike-sharing trips based on the POI feature vectors of their drop-off points. To determine the optimal K value, we test values ranging from 2 to 7 and calculate the corresponding silhouette coefficients. The results indicate that the clustering algorithm performs best when $K = 5$. Therefore, all bike-sharing trips are divided into five distinct types. Table 3.2 presents the clustering centres of the five categories, which reflect the distribution characteristics of POIs at the destinations of different trip types. The destinations of type A trips are predominantly associated with residential areas. Type B trips end in areas featuring shopping, dining, and service-related POIs. Type C trips are transport-related. Type D trips are linked to 'other' types of POIs. Type E trips have destinations surrounded by educational POIs. Figure 3.6 illustrates the spatial distribution of these five trip categories.

To validate the reliability of the clustering results, a comparison is made between the trip clustering results and the Gaode Satellite Map. Several typical areas are selected for comparison, as shown in Fig. 3.7. One of the gathering points for type A trips is confirmed to be the residential area of Oriole New Village. The large

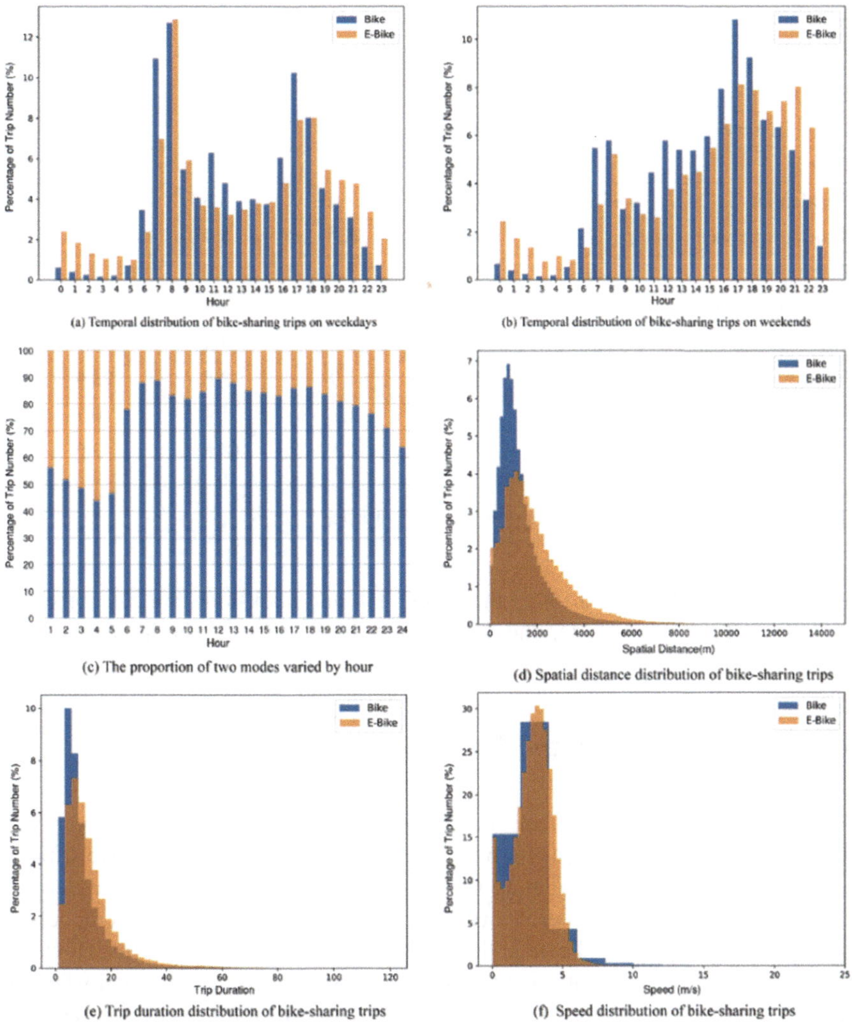

Fig. 3.5 Descriptive results of shared micromobility trips

commercial complex Tianyi Square is surrounded by type B trips, but no trip records are found within the square's boundaries, as it is designated as a pedestrian plaza. Type C trips generally end at intersections or along roads where metro stations and bus stops are commonly located. The figure also shows that some D-type trips end around parks, likely because the relatively small number of park POIs made them harder to identify using the K-means clustering, ultimately being labelled as 'other'. Additionally, trips on the Ningbo University campus are accurately classified as Type E trips. Interestingly, trips ending at the student dormitories are classified as Type A,

Table 3.2 The clustering centres of five trip categories

	Type A	Type B	Type C	Type D	Type E
Shopping	5.1	**20.1**	2.2	3.0	1.3
Daily life service	5.5	**20.7**	6.7	4.6	4.1
Eating	3.4	**17.4**	3.1	2.1	2.5
Residential	**65.9**	7.1	3.0	3.6	9.6
Transport services	4.4	6.3	**58.7**	5.7	0.4
Enterprises	1.0	7.2	3.3	3.0	0.7
Medical service	0.3	2.8	1.2	1.1	0.3
Public institution	1.8	2.7	2.1	3.4	2.9
Education	1.9	3.5	1.9	3.6	**73.9**
Recreation	0.7	2.8	1.2	1.4	0.7
Others	9.8	9.2	16.5	**68.6**	3.6

Note: The unit for all values is %.
The number in bold is the largest proportion of trip purposes

while trips with destinations near the dining hall and bookstores are categorised as Type B, further demonstrating the reliability of the K-means clustering method.

3.4.3 Travel Patterns and Corresponding Trip Purposes

We divide the corpus of trip labels into four sub-datasets: (a) bike trips on weekdays, (b) bike trips on weekends, (c) e-bike trips on weekdays, and (d) e-bike trips on weekends. We start by calculating the perplexity score for each sub-dataset (see Fig. 3.8). Generally, a lower perplexity score indicates a better-fitting LDA model that can effectively identify and extract the underlying structure of the corpus. However, an excessively low perplexity score may result from overfitting the model when the value of K exceeds a certain threshold. The optimal value of K typically corresponds to the turning point of the topic-perplexity curve. Based on the chart, we select the optimal K values for each sub-dataset: 6 for bike trips on weekdays, 8 for bike trips on weekends, 9 for e-bike trips on weekdays, and 5 for e-bike trips on weekends. Since the LDA model relies on word-frequency statistics, a higher K value may lead to duplications in travel patterns, while a lower K value could overlook relatively low-frequency travel patterns. By adopting distinct K values for each dataset, we are able to observe how people exhibit different travel patterns in varying micromobility contexts.

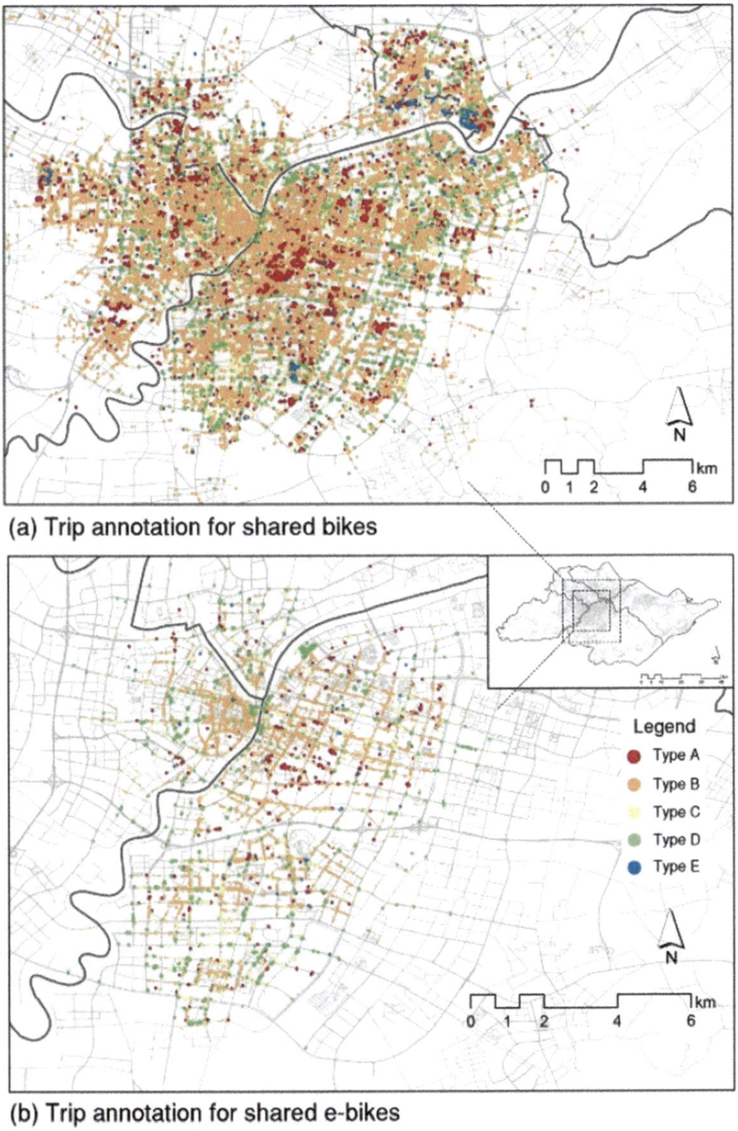

Fig. 3.6 Spatial distributions of five trip categories

3.4.3.1 Bike Travel Patterns

14 structured travel patterns are extracted from regular bike trips, with travel patterns 1–6 being weekdays, and patterns 7–14 being weekends, as shown in Fig. 3.9. Here, some representative patterns will be discussed.

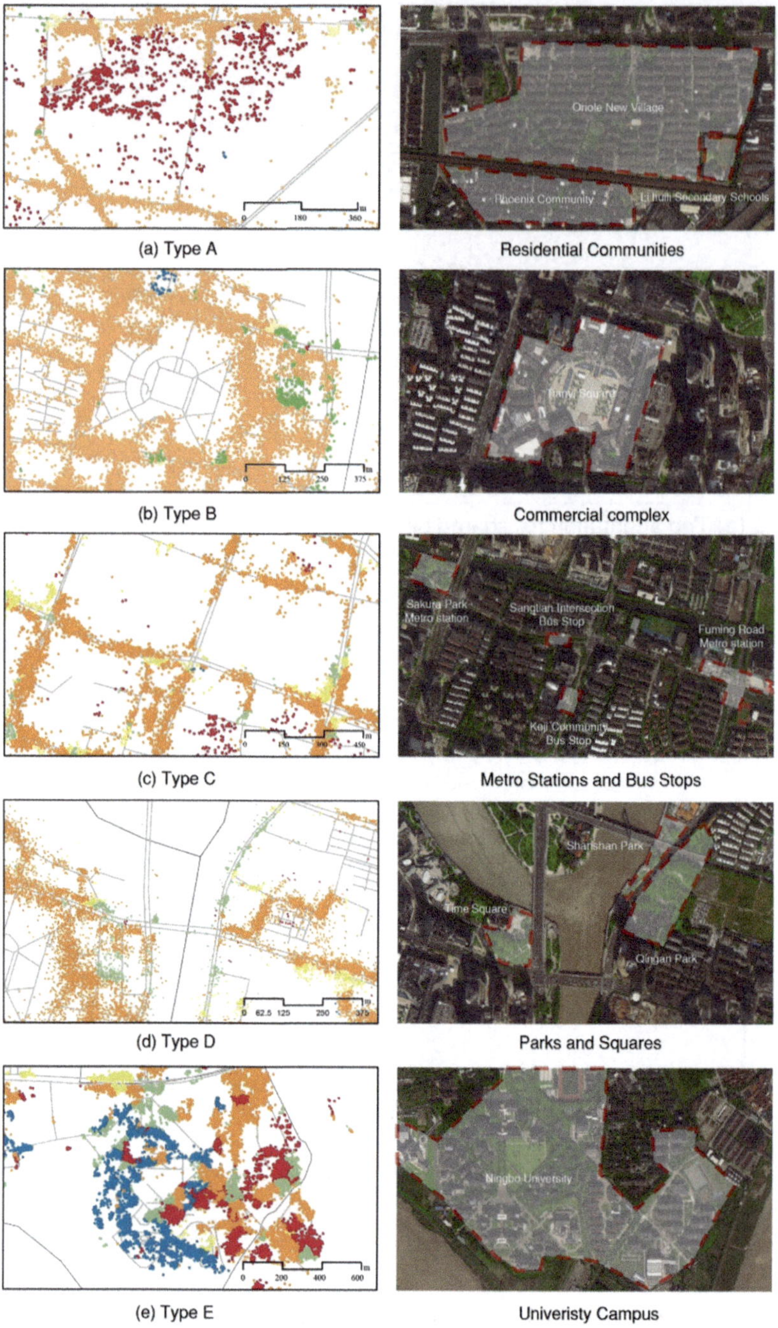

Fig. 3.7 The comparison of the trip clustering results with the Gaode Satellite Map

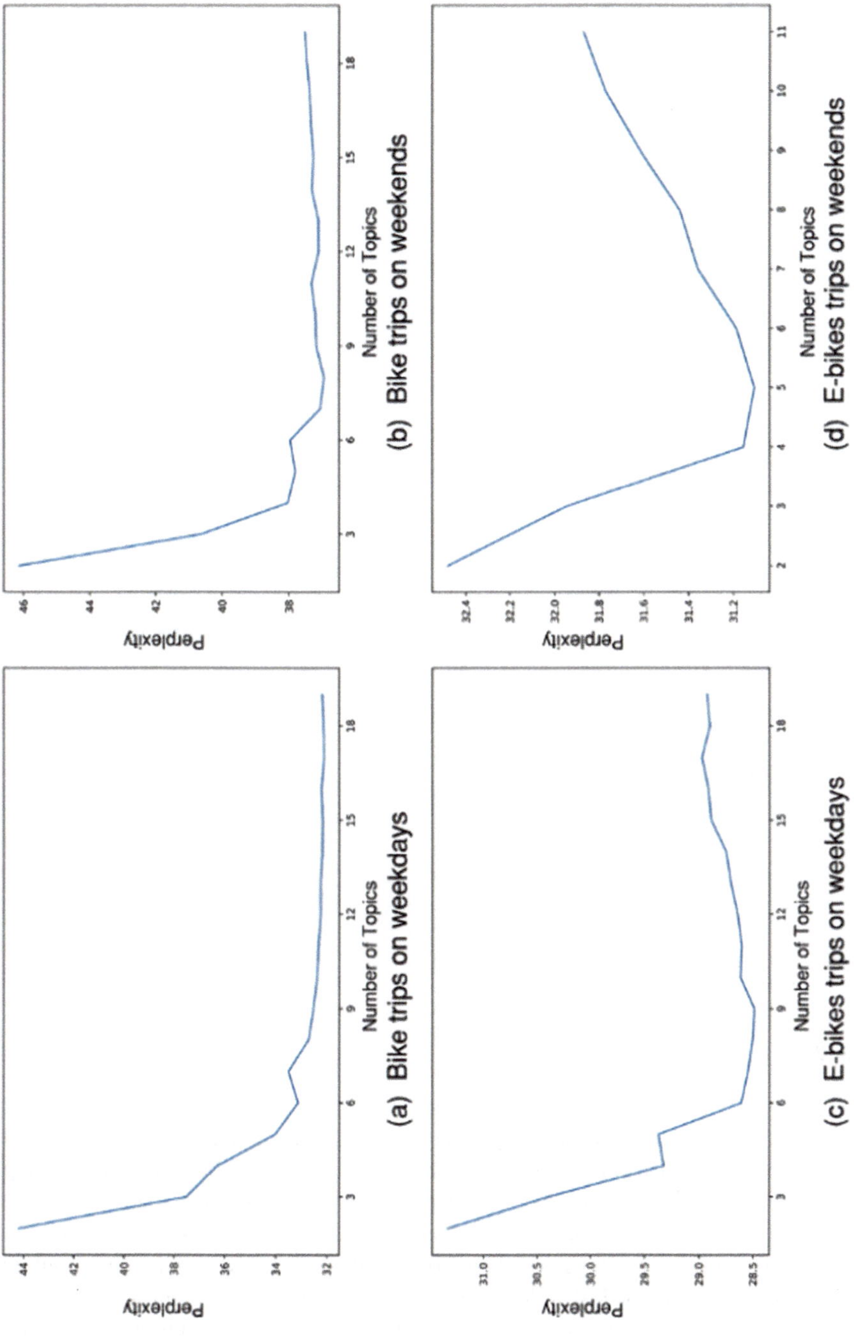

Fig. 3.8 Perplexities of four sub-datasets

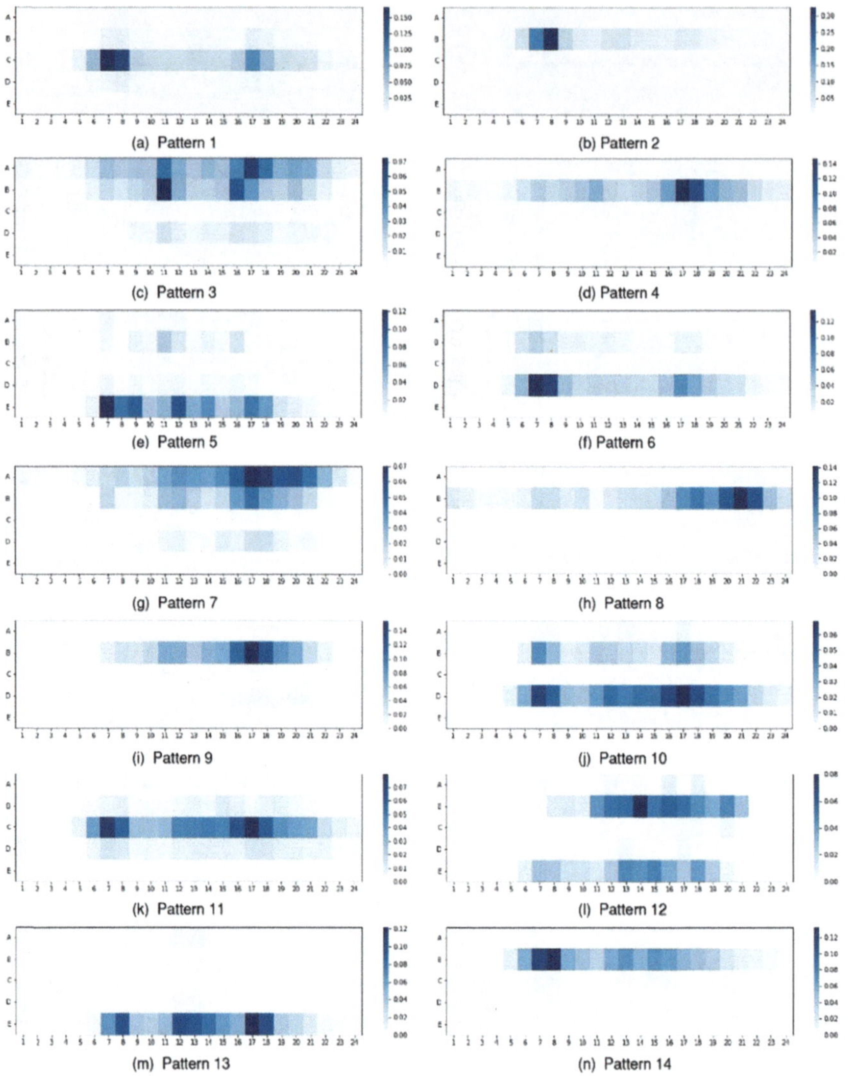

Fig. 3.9 Travel patterns of shared bike trips

Pattern 1 reveals that residents typically travel to transport-related POIs at 7 and 8 am, and at 5 and 6 pm. Parking lots, metro stations, and bus stops are among the most frequently visited transport-related POIs, indicating that shared bikes are often essential for users to access their desired modes of transport for commuting. This supports earlier findings that shared bikes can serve as a feeder mode for public transport, addressing the last-mile problem [44]. An additional insight is that parking lots also significantly contribute to the usage of shared bikes, a factor often overlooked in previous studies. Furthermore, more transport-related bike trips are made during the morning peak than the evening peak, aligning with Ma et al. [30], who suggested that travellers are more likely to choose shared bikes in the morning to avoid traffic jams and save time during rush hours.

Patterns 2 and 4 highlight numerous bike trips to service, eating, and shopping POIs around 8 am and 5 pm. According to Ross-Perez et al. [37], bike trips to retail stores or restaurants in the morning are typically for work-related purposes, whereas trips later in the day are usually for dining or shopping. Therefore, trip purposes can be inferred based on the arrival timestamp. However, a limitation to note is that some trips at 8 am may be for breakfast, which cannot be easily distinguished.

Pattern 3 illustrates that cyclists often travel to residential POIs at 11 am and between 4 and 6 pm. This finding aligns with the research by Xing et al.[43], who noted that bike-sharing trips to residential areas experience two rush hours: a midday mini-peak and an evening peak.

Pattern 5 predominantly occurs in areas with educational POIs, such as Ningbo University and the University of Nottingham Ningbo. The hourly distribution of trips is relatively even throughout the day, with three notable peaks: during breakfast and morning classes (7 am–9 am), at lunch (12 pm), and at dinner (6 pm). This aligns with Chicco and Diana's [13] study, which found that the temporal usage peaks of bike-sharing trips around universities correspond to the typical student schedule on campus.

Pattern 6 is most common on major roads with high traffic volumes. Its temporal distribution aligns with the typical commuting routine, exhibiting conventional morning and evening peaks. Determining specific purposes for these trips is challenging due to two factors: (1) the destination POIs, such as parks, are relatively few in number, making them difficult to identify using the K-means clustering algorithm, and (2) the 50 m destination buffer is relatively small, limiting the ability to accurately infer the true trip purposes. As a result, this pattern is classified under the 'other' category.

The travel patterns of bikes on weekends exhibit similarities to those on weekdays, but several notable differences are worth exploring. Pattern 7 mirrors Pattern 3, which can be interpreted as a lodging-related purpose. However, two distinctions emerge: First, the midday peak observed on weekdays disappears on weekends. Second, the evening arrival peak on weekends is both higher and occurs later than on weekdays. This shift in temporal distribution reflects real-world behaviour, suggesting that people are not required to ride bikes home for lunch or a midday break, as they might on working days, and are more likely to stay out later on weekends.

Pattern 11 is similar to Pattern 1 in terms of the travel purpose, but one key difference stands out: the evening peak on weekends is higher than the morning peak, whereas the reverse is true on weekdays. This shift is likely due to the fact that, on weekends, most people do not commute in the morning and are more inclined to go out for dinner and leisure activities in the evening.

Pattern 12 introduces a new trend, occurring in areas with educational POIs surrounded by shopping, eating, and service POIs. This pattern shows that bike users prefer to visit shops and obtain services around 2 pm on weekends, likely reflecting students' recreational activities during weekend afternoons.

3.4.3.2 E-Bike Travel Patterns

A total of 14 travel patterns for dockless e-bikes are identified, with 10 occurring on weekdays and 5 on weekends, as shown in Fig. 3.10.

Pattern 2 highlights that the peak arrival time of e-bikes at type B POIs is around 5 pm, which is strongly linked to dining and shopping purposes. Additionally, a significant number of e-bike users make trips from 7 pm to 12 am, suggesting that e-bikes are a popular mode of transport during the night. This aligns with findings

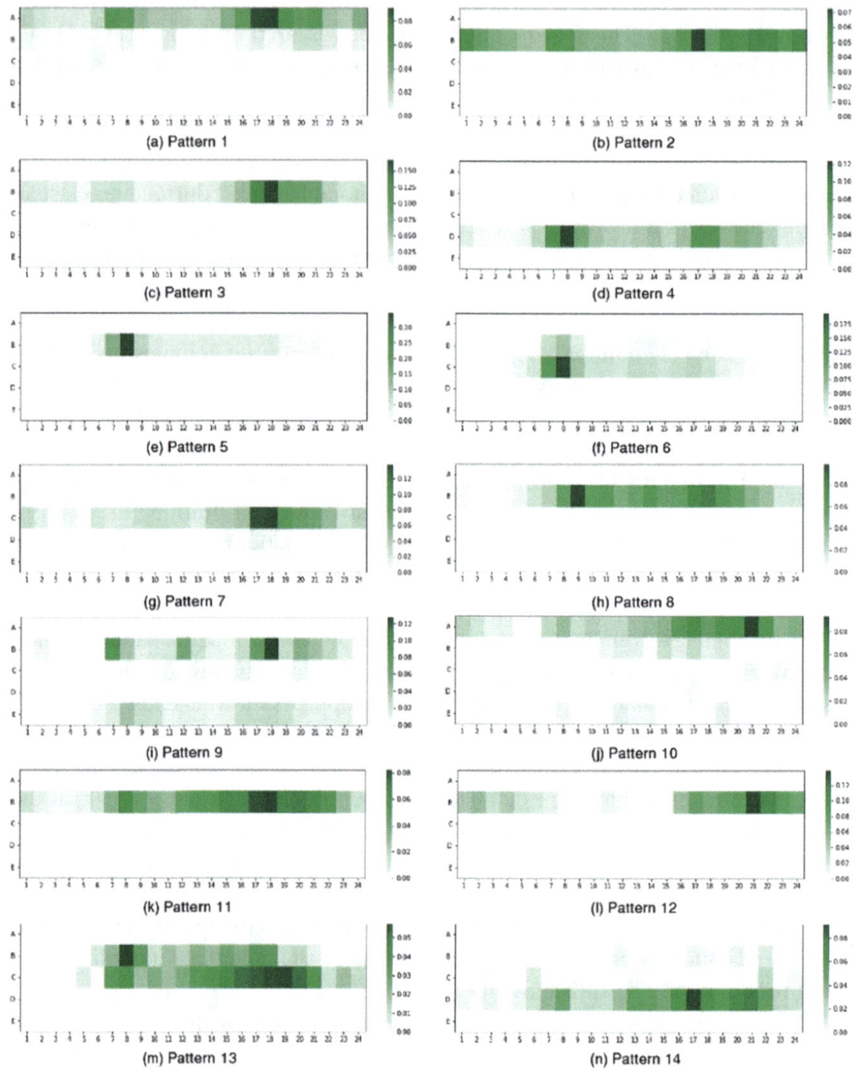

Fig. 3.10 Travel patterns of shared e-bike trips

from Almannaa et al. [3], who discovered consistent, low-speed e-bike usage between midnight and 11 am in Austin. A possible explanation for this trend is that e-bikes, with their higher speeds, enable users to reach their destinations more quickly, thus alleviating concerns about travel safety at night compared to regular bikes.

Patterns 6 and 7 show that people tend to ride e-bikes to transport facilities, such as public transport nodes, during both the morning rush hour and the evening peak. Notably, there are consistent e-bike trips occurring at midnight, which contrasts with the relatively few bike trips made at that time. This temporal pattern further highlights the importance of e-bikes for night travel.

Pattern 9, however, is less prominent compared to the other patterns. It appears in areas with POIs related to education, shopping, eating, and services. While individuals ride e-bikes to catering places, such as dining halls near educational institutions, around 6 pm, the number of trips falling into this pattern is relatively low compared to the number of shared bike trips. One possible reason for this is that e-bikes are less commonly used on campus due to safety concerns, while shared bikes remain predominant in student life. In fact, Jiageng et al. [21] recently noted that the prevalence of e-bikes on campuses requires coordination among various stakeholders, including students, school officials, and micromobility operators, to ensure safety and efficient use.

The e-bike travel patterns on weekends are quite similar to those observed on weekdays, with some notable differences. Pattern 10 is often seen in residential areas, but the evening peak (around 9 pm) occurs later than the typical weekday evening peak, as people tend to stay out and return home later on weekends. When compared to bike trips, it is evident that bike trips with the purpose of lodging tend to cluster between 4 and 7 pm, while e-bike trips reach a peak at 9 pm. This reinforces the idea that e-bikes are an essential mode of transport for night-time travel.

Patterns 11 and 12 illustrate that residents routinely travel to places for eating and shopping during the evening, with two significant peaks at dinner time (5–6 pm) and later at 9 pm. These patterns suggest a vibrant night-time economy in Ningbo.

Patterns 13 and 14 are associated with transport and other purposes, respectively. In comparison to weekday patterns with the same purposes, e-bike trips on weekends are more concentrated during the later hours of the day. This aligns with the weekend habits of residents, who tend to go out in the afternoon and evening, further emphasising the role of e-bikes in supporting evening and late-night mobility.

3.4.3.3 Aggregated Comparison of Trip Purposes

The trip purposes derived from the structured travel patterns indicate seven typical trip purposes for SMSs: transport, work, lodging, eating, shopping, education, and others. These are summarised in Table 3.3. The analysis reveals that eating is the most common trip purpose for both dockless bikes and e-bikes, accounting for 26.9% and 35.0% of all trips, respectively. This finding is consistent with previous studies [6, 26, 43], which also highlighted eating as a major purpose.

Table 3.3 Percentages of different trip purposes for shared bikes and e-bikes

Trip purposes	Shared bikes (%)	Shared e-bikes (%)
Transport	7.8	7.0
Work	17.5	17.3
Lodging	14.8	2.4
Eating	26.9	35.0
Shopping	16.7	32.0
Education	7.5	0.7
Others	8.7	5.5

For shared bikes, work is the second most common purpose, comprising 17.5% of all trips, closely followed by shopping at 16.7%. In contrast, for e-bikes, shopping accounts for a much larger share at 35.0%, while work makes up 17.3%. This suggests that e-bikes are particularly popular for shopping trips, while shared bikes are more often used for work-related trips. A potential explanation for this difference is related to the distinct infrastructure design for each type of shared micromobility systems. E-bikes may have more defined operational areas that align closely with commercial districts, which include eating and shopping POIs. Shared bikes, on the other hand, may not be as focused on these specific areas, which could explain the differences in the trip purpose distribution.

The lodging purpose, which refers to trips to and from home, accounts for approximately 14.8% of bike trips, which highlights its significance. However, this purpose is much less common for e-bikes, representing only 2.4% of all e-bike trips. This striking difference challenges some earlier assumptions that shared e-bikes play a dominant role in residential travel. A possible explanation for this finding is that e-bikes are less convenient than shared bikes for lodging-related trips due to parking restrictions. E-bikes often need to be parked within designated electronic geofenced areas, which are more limited in scope compared to the more flexible parking options for regular bikes. This stricter regulation, primarily driven by power management and safety concerns, leads to fewer e-bike parking stations around residential areas. As a result, the use of shared e-bikes for trips to and from home is much less prevalent compared to regular bikes, where parking is generally more accessible and flexible.

The percentage of bike and e-bike trips for transport purposes in Ningbo is 7.8% and 7.0%, respectively. This suggests that a notable portion of residents rely on bike-sharing as a feeder mode to connect to public transit or to access parking lots for their private cars. This finding aligns with studies by Zhou et al. [13] and Zhou et al. [50], which highlighted the role of bike-sharing in facilitating access to other transport modes. However, the proportion of transport-related trips in Ningbo is lower than the 19.58% estimated by Xing et al. [43] for Shanghai, a much larger and more transit-oriented city. The discrepancy could be explained by the difference in the city scale: Shanghai, as a megacity, has a well-established metro system that is heavily relied upon for long-distance commuting, thus increasing the proportion of transport-related trips. In contrast, Ningbo, with a smaller city scale and a more diverse range of commuting options, sees relatively fewer bike-sharing trips serving as a feeder to public transport.

Bike-sharing trips related to educational facilities have been documented in previous studies [2, 21]. In this study, trips with an education purpose account for 7.5% of all bike trips, but only 0.7% of all e-bike trips. This significant discrepancy likely arises from the restrictions on e-bike usage on university campuses, particularly for safety reasons, as universities may limit the operation of e-bikes to reduce traffic-related risks for students. In contrast, in countries like the United States, Singapore, and Italy, there are fewer restrictions on the use of electric SMSs on university campuses. Studies in these countries have observed high volumes of e-scooter traffic in and around educational areas [5, 13, 51]. This suggests that the regulatory environment significantly impacts the use of e-bikes on campuses, and more relaxed regulations in other countries may facilitate greater adoption of electric SMSs for educational trips.

3.4.3.4 Spatial Distribution of Trip Purposes

Figure 3.11 illustrates the spatial distribution of different travel purposes for both shared bike and e-bike trips. For shared bike trips, the lodging purpose is concentrated in certain grids, particularly near large residential communities like Green Lotus Court and Oriole New Village. This suggests that the shared bike infrastructure is well-developed in these areas, and residents are likely to rely on the bike-sharing system for commuting. In contrast, e-bike trips related to lodging are relatively few. Only a few residential areas, such as Ningdong Home and Ruyi Water Bay, show any significant e-bike usage for lodging purposes. This indicates that while e-bike infrastructure is available in these residential communities, its coverage is not as widespread as that of shared bikes. Therefore, future e-bike planning should focus on expanding the network to areas that lack sufficient e-bike service coverage, particularly in other residential communities where shared e-bike services are currently not available.

Bike trips with the purpose of education are commonly observed on university campuses, such as Ningbo University and The University of Nottingham Ningbo, reflecting the frequent use of bike-sharing for commuting between academic buildings and campus facilities. On the other hand, e-bike trips with an educational purpose are more prevalent around primary and secondary schools, such as Depei Primary School. This pattern suggests that parents tend to use e-bikes for school drop-offs, likely due to the time-saving advantage e-bikes offer. This trend aligns with findings from Zhou et al. [50], who noted that e-bikes are commonly used for short-distance travel, particularly in situations where convenience and speed are priorities, such as taking children to school.

The hotspots for work-related bike and e-bike trips show significant overlap, typically occurring in areas that attract enterprises, industrial parks, and commercial complexes, which are potential workplaces for residents. These areas see high

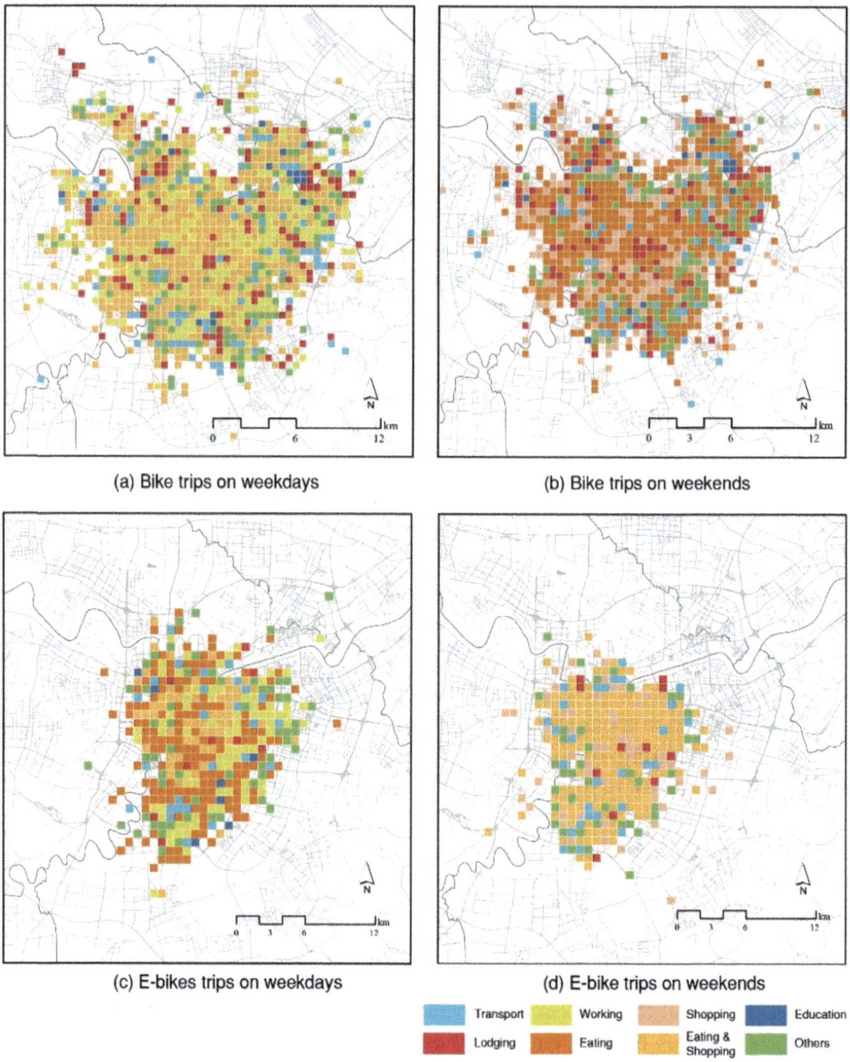

Fig. 3.11 Spatial patterns of trip purposes

concentrations of work-related trips, reflecting the commuting patterns of individuals heading to their jobs. Notably, work-related trips are absent on weekends, as the assumption is made that people do not work during that time. However, this may be an area for future research, considering that a significant number of people work on weekends, especially in industries or sectors with non-traditional working hours, which could impact the demand for bike-sharing services on weekends.

The distribution of transport-related trips aligns with the locations of public transport hubs, including metro stations, bus stops, and parking lots. This reflects the role of shared bikes and e-bikes as complementary modes for accessing public transport or private vehicles, reinforcing the idea that bike-sharing systems serve as a feeder mode for commuters. Trips with the purpose of eating and shopping are more dispersed across Ningbo, indicating that these types of trips are not confined to a specific area but rather spread throughout the city, likely near retail hubs and dining districts. On the other hand, trips categorised under other purposes tend to cluster in areas with high traffic volumes, such as major roads, parks, and open spaces.

3.5 Conclusions and Discussion

3.5.1 Main Conclusions

This chapter provides valuable insights into the travel patterns and trip purposes of shared bikes and e-bikes within a dockless bike-sharing system, analysed through an LDA-based framework. Based on the study, several key conclusions can be made:

Short-distance travel and mode characteristics: Both shared bikes and e-bikes play a crucial role in short-distance travel, with e-bikes covering longer distances and longer durations compared to regular bikes. The speed distribution of e-bikes is more varied, reflecting their reliance on electric power, which helps users maintain diverse speeds and conserve energy. In contrast, bike speeds tend to be more consistent and slightly left-skewed, indicating more stable travel behaviour.

Travel pattern consistency and night-time usage: The study observed familiar morning and evening rush hours during weekdays, with peaks in the afternoon and evening on weekends. Notably, shared e-bikes are found to be more popular for night-time travel than traditional bikes, which aligns with earlier studies that highlight e-bikes as more suited for late-night journeys due to their convenience and speed.

Typical trip purposes: This chapter identifies seven typical trip purposes: transport, work, lodging, eating, shopping, education, and others. Among these, eating is the most common purpose for both shared bikes and e-bikes, with transport-related trips being less significant than anticipated. Interestingly, lodging and education-related trips are more common for shared bikes, while e-bikes have fewer trips associated with these purposes, possibly due to parking restrictions and safety concerns around campus areas for e-bikes.

Differences in trip purposes: The study also points out some critical differences between shared bikes and e-bikes. For example, e-bikes are used less frequently for educational and lodging purposes compared to shared bikes. This might be related to operational constraints of e-bikes, including geofencing and parking limitations, which restrict their use in certain areas like residential neighbourhoods and campuses.

Overall, the analysis not only contributes to understanding the spatial and temporal dynamics of dockless bike-sharing systems but also reveals nuanced differences between e-bikes and shared bikes in terms of usage patterns and trip purposes. This provides valuable insights for urban planners and transport researchers aiming to optimise micromobility systems.

3.5.2 Implications for Urban and Transport Planning Practices

Based on the findings, some targeted planning and regulation suggestions can be given.

For the SMSs' operators, they can adopt appropriate operation strategies:

(1) Surplus bike and e-bike retrieval in commercial areas: Given that eating and shopping are the most common trip purposes for shared bikes and e-bikes, many bikes are left in commercial areas after evening activities, particularly on weekends. This results in an over-concentration of fleets in these areas. Tianyi Square, for instance, is identified as an ideal location for fleet retrieval after the evening peak. By reallocating surplus bikes and e-bikes from commercial areas to nearby residential neighbourhoods, operators can better support the morning commute—as residential areas are the origins of most daily trip chains. This strategy would enhance the system's efficiency and ensure that bikes are available where they are most needed, improving overall service quality and user satisfaction.

(2) Increase bike and e-bike supply and expand parking areas: To meet demand, shared bike and e-bike supply should be increased, particularly around residential areas. Bike-sharing companies could focus on redistributing shared bikes before the morning peak, ensuring they are readily available to users during the most common commuting hours. For e-bikes, the proportion of home-related trips is surprisingly low, possibly due to insufficient geofencing areas that limit where e-bikes can be parked. To address this, more designated e-bike parking zones should be established in residential neighbourhoods. This would make it easier for users to access e-bikes for commuting and short-distance travel, improving the overall effectiveness of the bike-sharing system and encouraging greater use in residential areas where the demand for e-bikes is high.

(3) Implement dynamic rebalancing strategies for transport-related POIs: Dynamic rebalancing can be used to redistribute bikes and e-bikes around transport-related POIs, such as public transport stations and parking lots, to address the first and last-mile problem. SMSs are commonly used as commuter feeder modes, making it essential for bike fleets to align with commuter flow patterns. In the morning, there is often an excess of bikes near residential communities (as people use them to reach transport stations) and a shortage around workplaces. The same issue occurs in the evening peak hours, but in reverse. To address

this, operators can transfer bikes between metro and bus stations near residential areas and workplaces to ensure a demand–supply balance. Additionally, parking lots are often overlooked during fleet rescheduling. Operators should also rebalance fleets in parking areas during peak times to ensure adequate supply in these high-demand locations. By implementing these strategies, the overall efficiency and availability of the bike-sharing system will improve, especially during peak commuting times.

(4) Support shared e-bike usage on university campuses: To promote the use of shared e-bikes on college campuses, supporting facilities and equipment should be provided. While university students are among the primary users of SMSs, e-bike usage in educational institutions remains limited compared to shared bikes. However, studies in countries like the Netherlands and China show that private e-bike usage is increasing on university campuses due to the longer distances between buildings, which makes e-bikes a more efficient mode of transport. The slow development of shared e-bike infrastructure and safety concerns are the primary barriers to the wider adoption of shared e-bikes on campuses. Campuses with well-developed bike-sharing infrastructure have significant growth potential for shared e-bikes. To unlock this potential, operators should deploy more shared e-bikes on university campuses and provide necessary safety equipment, such as cycling helmets. Additionally, speed limiters should be installed on e-bikes to address safety concerns. These actions will help encourage students to choose shared e-bikes as a preferred mode of transport, enhancing the overall mobility options on campuses.

For the government, it can implement appropriate urban and transport planning strategies to efficiently utilise the SMSs in Ningbo:

(1) Develop rational land use planning to optimise SMS infrastructure: To enhance the integration and utilisation of the SMS infrastructure, rational land use planning should be developed based on the extracted travel patterns of bikes and e-bikes. For instance, commercial areas exhibit a high concentration of idle SMS fleets at night, which suggests that these surplus fleets should be reallocated to nearby residential areas to meet the demand during peak hours. One way to address this challenge is through mixed land-use development, specifically focusing on high-density commercial and residential areas. In the case of Ningbo, planning new urban development with a mix of commercial and residential spaces would reduce the need for extensive fleet rescheduling across different urban zones. This strategic land use could optimise fleet utilisation, creating an internal cycle where supply and demand for SMS infrastructure are balanced within a specific area. By integrating land use with mobility patterns, urban planners can improve the efficiency of SMS operations, minimise the environmental impact of fleet redistribution, and promote sustainable urban development that supports both residents and businesses.

(2) Encourage multimodal transport by integrating SMSs into the current public transport landscape. Although studies have shown that SMSs, especially shared e-bikes, may compete with and substitute public transport, the government can

encourage the deployment and operation of SMSs to complement transport services in areas not covered by buses or metros, thereby meeting the transport needs of residents in the whole region. Additionally, SMSs can serve as feeder modes of public transport to solve the first/last mile problem. Therefore, the government can promote the integration of SMSs and public transport by planning more cycling road networks and SMS parking areas next to bus or metro stations to spur the prosperity of multimodal transport in the city.

3.5.3 Limitations and Future Directions

This chapter has some limitations that should be addressed in future research. First, the methodology for trip annotation could be further optimised. In this study, trips are labelled based on the K-means clustering algorithm, using POI feature vectors. As a result, the number of POIs might have influenced the final imputation of trip purposes. The inferred purposes are often related to functions associated with a larger number of POIs, while niche travel reasons, such as medical purposes, are difficult to identify. Therefore, future work could enhance the trip annotation method to account for functions represented by a smaller number of POIs. Second, limitations may also exist in the datasets used. The limited time span of the travel dataset (just one week) may lead to conclusions that are not universally applicable or robust. Additionally, the trip dataset is derived solely from the operator Hellobike, whereas there are other smaller micromobility operators in Ningbo. The operational data from only the leading operator may be incomplete. Therefore, future studies are encouraged to expand the dataset by including a longer time frame and data from more micromobility operators.

This work also provides several directional implications for future studies. The first direction is to expand the research scope. Specifically, we only investigated the trip purposes of shared bikes and e-bikes due to the limitations of the dataset. Future research could select a city with a variety of SMSs, such as New York, to include other SMSs like shared e-scooters and skateboards, thus offering a more holistic view of the SMS usage. Another direction involves considering the pick-up point of each trip in the trip annotation. By incorporating information about pick-up and drop-off points into trip labels, the LDA model could extract structural travel patterns between different origin–destination pairs. These trip chains would provide valuable insights into travel demand between specific origin–destination pairs. New research could also explore the deeper logic behind trip-purpose differences across different SMSs. Factors influencing the use of SMSs, such as the built environment, could also be investigated. For example, by examining the characteristics of hotspots for lodging-related trips, it may become apparent that the lack of bike lanes or insufficient parking areas contributes to a low number of lodging-related trips. Consequently, urban planners and operators could implement targeted improvements to the urban environment and shared infrastructure planning around residential communities.

References

1. Abduljabbar RL, Liyanage S, Dia H (2021) The role of micro-mobility in shaping sustainable cities: a systematic literature review. Transp Res Part Transp Environ 92:102734. https://doi.org/10.1016/j.trd.2021.102734
2. Aliari S, Nasri A, Motalleb Nejad M, Haghani A (2020) Toward sustainable travel: an analysis of campus bikeshare use. Transp Res Interdiscip Perspect 6:100162. https://doi.org/10.1016/j.trip.2020.100162
3. Almannaa MH, Ashqar HI, Elhenawy M et al (2021) A comparative analysis of e-scooter and e-bike usage patterns: findings from the City of Austin, TX. Int J Sustain Transp 15:571–579. https://doi.org/10.1080/15568318.2020.1833117
4. Alsger A, Tavassoli A, Mesbah M et al (2018) Public transport trip purpose inference using smart card fare data. Transp Res Part C Emerg Technol 87:123–137. https://doi.org/10.1016/j.trc.2017.12.016
5. Bai S, Jiao J (2020) Dockless E-scooter usage patterns and urban built Environments: a comparison study of Austin, TX, and Minneapolis, MN. Travel Behav Soc 20:264–272. https://doi.org/10.1016/j.tbs.2020.04.005
6. Bao J, Xu C, Liu P, Wang W (2017) Exploring bikesharing travel patterns and trip purposes using smart card data and online point of interests. Netw Spat Econ 17:1231–1253. https://doi.org/10.1007/s11067-017-9366-x
7. Bieliński T, Kwapisz A, Ważna A (2021) Electric bike-sharing services mode substitution for driving, public transit, and cycling. Transp Res Part Transp Environ 96:102883. https://doi.org/10.1016/j.trd.2021.102883
8. Bieliński T, Ważna A (2020) Electric scooter sharing and bike sharing user behaviour and characteristics. Sustainability 12:1–13
9. Blei DM, Ng AY, Jordan MI (2003) Latent Dirichlet allocation. J Mach Learn Res 3:993–1022
10. Caspi O (2022) Equity implications of electric bikesharing in Philadelphia. GeoJournal. https://doi.org/10.1007/s10708-022-10698-1
11. Chang X, Wu J, He Z et al (2020) Understanding user's travel behavior and city region functions from station-free shared bike usage data. Transp Res Part F Traffic Psychol Behav 72:81–95. https://doi.org/10.1016/j.trf.2020.03.018
12. Chen Q, Pan X, Liu F et al (2022) Reposition optimization in free-floating bike-sharing system: a case study in Shenzhen City. Phys Stat Mech Its Appl 593:126925. https://doi.org/10.1016/j.physa.2022.126925
13. Chicco A, Diana M (2022) Understanding micro-mobility usage patterns: a preliminary comparison between dockless bike sharing and e-scooters in the city of Turin (Italy). Transp Res Procedia 62:459–466. https://doi.org/10.1016/j.trpro.2022.02.057
14. Ermagun A, Fan Y, Wolfson J et al (2017) Real-time trip purpose prediction using online location-based search and discovery services. Transp Res Part C Emerg Technol 77:96–112. https://doi.org/10.1016/j.trc.2017.01.020
15. Gong L, Liu X, Wu L, Liu Y (2016) Inferring trip purposes and uncovering travel patterns from taxi trajectory data. Cartogr Geogr Inf Sci 43:103–114. https://doi.org/10.1080/15230406.2015.1014424
16. Guidon S, Becker H, Dediu H, Axhausen KW (2019) Electric bicycle-sharing: a new competitor in the urban transportation market? An empirical analysis of transaction data. Transp Res Rec J Transp Res Board 2673:15–26. https://doi.org/10.1177/0361198119836762
17. Han F, Xie R, Lai M (2018) Traffic density, congestion externalities, and urbanization in China. Spat Econ Anal 13:400–421. https://doi.org/10.1080/17421772.2018.1459045
18. Han G, Sohn K (2016) Activity imputation for trip-chains elicited from smart-card data using a continuous hidden Markov model. Transp Res Part B Methodol 83:121–135. https://doi.org/10.1016/j.trb.2015.11.015
19. Han SS (2020) The spatial spread of dockless bike-sharing programs among Chinese cities. J Transp Geogr 86:102782. https://doi.org/10.1016/j.jtrangeo.2020.102782

20. He Y, Song Z, Liu Z, Sze NN (2019) Factors influencing electric bike share ridership: analysis of Park City, Utah. Transp Res Rec 2673:12–22. https://doi.org/10.1177/0361198119838981
21. Jiageng N, Lanlan Z, Xianghong L (2022) A study on the trip behavior of shared bicycles and shared electric bikes in Chinese universities based on NL model—Henan Polytechnic University as an example. Phys Stat Mech Appl 604:127855. https://doi.org/10.1016/j.physa.2022.127855
22. Kazemzadeh K, Ronchi E (2022) From bike to electric bike level-of-service. Transp Rev 42:6–31. https://doi.org/10.1080/01441647.2021.1900450
23. Kou Z, Cai H (2019) Understanding bike sharing travel patterns: An analysis of trip data from eight cities. Phys Stat Mech Appl 515:785–797. https://doi.org/10.1016/j.physa.2018.09.123
24. Krause CM, Zhang L (2019) Short-term travel behavior prediction with GPS, land use, and point of interest data. Transp Res Part B Methodol 123:349–361. https://doi.org/10.1016/j.trb.2018.06.012
25. Kroesen M (2017) To what extent do e-bikes substitute travel by other modes? Evidence from the Netherlands. Transp Res Part Transp Environ 53:377–387. https://doi.org/10.1016/j.trd.2017.04.036
26. Li A, Huang Y, Axhausen KW (2020) An approach to imputing destination activities for inclusion in measures of bicycle accessibility. J Transp Geogr 82:102566. https://doi.org/10.1016/j.jtrangeo.2019.102566
27. Li A, Zhao P, Huang Y et al (2020) An empirical analysis of dockless bike-sharing utilization and its explanatory factors: case study from Shanghai. China. J Transp Geogr 88:102828. https://doi.org/10.1016/j.jtrangeo.2020.102828
28. Li Y, Chen Q, Ma Q et al (2022) Injuries and risk factors associated with bicycle and electric bike use in China: a systematic review and meta-analysis. Saf Sci 152:105769. https://doi.org/10.1016/j.ssci.2022.105769
29. Liu Y (2022) A hybrid big-data-based and tolerance-based method to estimate environmental benefits of electric bike sharing. Appl Energy 315:118974
30. Ma X, Ji Y, Yuan Y et al (2020) A comparison in travel patterns and determinants of user demand between docked and dockless bike-sharing systems using multi-sourced data. Transp Res Part Policy Pract 139:148–173. https://doi.org/10.1016/j.tra.2020.06.022
31. Macqueen J (1967) Some methods for classification and analysis of multivariate observations. In: 5th Berkeley symposium on mathematical statistics and probability, pp 281–297
32. Madapur B, Madangopal S, Chandrashekar MN (2020) Micro-mobility infrastructure for redefining urban mobility. Eur J Eng Sci Technol 3:71–85. https://doi.org/10.33422/ejest.v3i1.163
33. McKenzie G (2019) Spatiotemporal comparative analysis of scooter-share and bike-share usage patterns in Washington, D.C. J Transp Geogr 78:19–28. https://doi.org/10.1016/j.jtrangeo.2019.05.007
34. Milakis D, Gedhardt L, Ehebrecht D, Lenz B (2020) Is micro-mobility sustainable? An overview of implications for accessibility, air pollution, safety, physical activity and subjective wellbeing. In: Curtis C (ed) Handbook of sustainable transport. Edward Elgar Publishing, Cheltenham, pp 180–189
35. Pettersson I, Hagberg L, Fredriksson C, Hermansson LN (2016) The effect of powered scooters on activity, participation and quality of life in elderly users. Disabil Rehabil Assist Technol 11:558–563. https://doi.org/10.3109/17483107.2015.1027301
36. Reck DJ, Haitao H, Guidon S, Axhausen KW (2021) Explaining shared micromobility usage, competition and mode choice by modelling empirical data from Zurich, Switzerland. Transp Res Part C Emerg Technol 124:102947. https://doi.org/10.1016/j.trc.2020.102947
37. Ross-Perez A, Walton N, Pinto N (2022) Identifying trip purpose from a dockless bike-sharing system in Manchester. J Transp Geogr 99:103293. https://doi.org/10.1016/j.jtrangeo.2022.103293
38. Sari Aslam N, Zhu D, Cheng T et al (2021) Semantic enrichment of secondary activities using smart card data and point of interests: a case study in London. Ann GIS 27:29–41. https://doi.org/10.1080/19475683.2020.1783359

39. Shaer A, Rezaei M, Moghani Rahimi B, Shaer F (2021) Examining the associations between perceived built environment and active travel, before and after the COVID-19 outbreak in Shiraz city, Iran. Cities 115:103255. https://doi.org/10.1016/j.cities.2021.103255
40. Shaheen SA, Martin EW, Cohen AP, Finson RS (2012) Public bikesharing in North America: early operator and user understanding. IEEE. https://doi.org/10.3141/2387-10
41. Wang Q (2018) Urbanization and global health: the role of air pollution. Iran J Public Health 47:1644–1652
42. Weis C, Kowald M, Danalet A, et al. (2021) Surveying and analysing mode and route choices in Switzerland 2010–2015. Travel Behaviour and Society 22: 10–21. https://doi.org/10.1016/j.tbs.2020.08.001.
43. Xing Y, Wang K, Lu JJ (2020) Exploring travel patterns and trip purposes of dockless bike-sharing by analyzing massive bike-sharing data in Shanghai, China. J Transp Geogr 87:102787. https://doi.org/10.1016/j.jtrangeo.2020.102787
44. Yang X-H, Cheng Z, Chen G et al (2018) The impact of a public bicycle-sharing system on urban public transport networks. Transp Res Part Policy Pract 107:246–256. https://doi.org/10.1016/j.tra.2017.10.017
45. Ying JJ-C, Lee W-C, Tseng VS (2014) Mining geographic-temporal-semantic patterns in trajectories for location prediction. ACM Trans Intell Syst Technol 5:2:1-2:33. https://doi.org/10.1145/2542182.2542184
46. Zhang Y, Brussel MJG, Thomas T, van Maarseveen MFAM (2018) Mining bike-sharing travel behavior data: an investigation into trip chains and transition activities. Comput Environ Urban Syst 69:39–50. https://doi.org/10.1016/j.compenvurbsys.2017.12.004
47. Zhang Y, Lin D, Liu XC (2019) Biking islands in cities: an analysis combining bike trajectory and percolation theory. J Transp Geogr 80:102497. https://doi.org/10.1016/j.jtrangeo.2019.102497
48. Zhang Y, Lin D, Mi Z (2019) Electric fence planning for dockless bike-sharing services. J Clean Prod 206:383–393. https://doi.org/10.1016/j.jclepro.2018.09.215
49. Zhang Y, Mi Z (2018) Environmental benefits of bike sharing: a big data-based analysis. Appl Energy 220:296–301. https://doi.org/10.1016/j.apenergy.2018.03.101
50. Zhou X, Ji Y, Yuan Y et al (2022) Spatiotemporal characteristics analysis of commuting by shared electric bike: a case study of Ningbo, China. J Clean Prod 362:132337. https://doi.org/10.1016/j.jclepro.2022.132337
51. Zhu R, Zhang X, Kondor D et al (2020) Understanding spatio-temporal heterogeneity of bike-sharing and scooter-sharing mobility. Comput Environ Urban Syst 81:101483. https://doi.org/10.1016/j.compenvurbsys.2020.101483
52. Zou Q, Yao X, Zhao P et al (2018) Detecting home location and trip purposes for cardholders by mining smart card transaction data in Beijing subway. Transportation 45:919–944. https://doi.org/10.1007/s11116-016-9756-9

Chapter 4
Evaluate Carbon Emission Reductions from Shared Micromobility

4.1 Introduction

Global warming and climate change have been long-standing global concerns. Greenhouse gases, such as carbon dioxide (CO_2), methane, and nitrous oxide, are the primary drivers of climate change and human livability. Among these, CO_2 emissions have the most significant impact, accounting for approximately 65% of total greenhouse gas emissions [44]. Since the transport sector is responsible for around a quarter of global greenhouse gas emissions [20], there is an urgent need to adopt more environmentally friendly modes of transport to help reduce carbon emissions and tackle the challenges posed by climate change [26].

Recently, shared micromobility has emerged as a significant development in green urban transport, referring to the shared use of low-speed, small, and lightweight vehicles that are available on an as-needed basis for travel. Three types of shared micromobility are widely adopted today: shared bikes, electric bikes (e-bikes), and electric scooters (e-scooters). This chapter focuses on the first two, as the primary concern is shared micromobility systems (SMSs) in China, where shared e-scooter systems are currently not permitted. SMSs have gained particular success with the advent of dockless services, offering greater convenience and flexibility: users can rent bikes or e-bikes via smartphones and park them within any designated geofence upon completing a trip [33, 48]. By 2030, the global shared micromobility market is projected to reach a value of $90 billion, representing an approximate 40% increase from 2019 [30].

Shared micromobility offers several advantages, including the reduction of traffic congestion, enhanced integration with public transit, promotion of a healthy lifestyle, provision of alternative travel options, and affordability with flexible usage [3, 16]. It is also widely regarded as having the potential to reduce carbon emissions by replacing trips made using motorised modes of transport such as private cars and taxis [38, 47]. To better understand how SMSs contribute to urban sustainability, a growing body of research has sought to quantify the environmental benefits associated with

Y. Zhang, *Understanding Shared Micromobility in China Using Massive User-Generated Trip Data*, Urban Sustainability,
https://doi.org/10.1007/978-981-96-8492-2_4

shared bikes [9, 28] and e-bikes [22, 27]. The primary approach for quantifying these benefits involves calculating the difference in carbon emissions between using shared micromobility and using other transport modes, assuming that shared micromobility was unavailable for the same trip. Despite existing studies offering practical methods, various cases, and valuable insights into the carbon reduction potential of SMSs, several research gaps remain that have yet to be fully addressed.

Firstly, the potential for carbon emission reductions from shared e-bikes has been infrequently explored, and comparative studies assessing the reduction potential of shared bikes and e-bikes are notably lacking. Shared e-bikes offer higher speeds and require less physical effort than human-powered shared bikes [12]. This makes them more appealing to users, as they allow for longer trips and enable riders to overcome road barriers with greater ease than shared bikes [37]. This could lead to a distinct substitution pattern between shared micromobility and alternative transport modes: shared bikes are more likely to replace short trips (e.g., walking), whereas shared e-bikes are more commonly used to substitute trips made by private cars and public transport [51]. Furthermore, e-bike sharing is less environmentally friendly than bike sharing, as it directly contributes to carbon emissions through electricity consumption. Considering these differences, the carbon reduction potential of shared e-bikes is likely to differ significantly from that of shared bikes. However, since e-bike sharing is a relatively new service, only a limited number of studies have examined its environmental benefits [27, 51], and even fewer have investigated the spatiotemporal heterogeneity of the carbon reduction potential between e-bike-sharing and bike-sharing.

Secondly, most of the previous literature has conducted case studies in a single city (occasionally in several cities, e.g., [22]), such as Edinburgh [9], New York City [6], Shanghai [8, 49] and Zurich [35]. The cities examined are primarily economically well-developed, either globally or within their respective countries, while medium- or small-sized cities in less mature stages of development have been largely overlooked. Although case studies from large cities have provided valuable insights into the potential for carbon reduction, they tend to neglect the variety of spatiotemporal patterns of mobility, carbon emissions, and urban contexts in cities at different developmental levels. Relying primarily on data from first-tier cities risks developing a limited understanding of the underlying processes and could lead to biased urban policy interventions [14, 31].

Thirdly, a crucial step in quantifying the environmental impacts of SMSs is to estimate how shared micromobility substitutes other transport modes [2, 23]. In many studies calculating environmental benefits, shared micromobility is often assumed to replace walking and driving [21, 49]. However, in reality, it also serves as a substitute for public transit, which is generally more environmentally friendly compared to private cars [17, 22, 38]. Failing to account for the potential substitution of buses and subways may lead to an overestimation of the carbon emissions reduced by shared micromobility. Although Saltykova et al. [38] and Li et al. [22] have considered the substitution of public transit in their environmental impact assessments of shared micromobility, neither study provides a comparative analysis between dockless shared bikes and e-bikes.

Last but not least, this study further explores the carbon emissions of trips with different purposes to gain a deeper understanding of the environmental impacts of SMSs. Existing studies have developed various methods to accurately infer trip purposes, which are often not recorded in user-generated trip data, to better understand users' motivations for using SMSs [43, 45]. However, the relationship between SMSs' carbon reduction potential and users' trip purposes has yet to be investigated. Users are motivated to use a bike or an e-bike for various activities (e.g., shopping) in different urban contexts. Trips driven by different purposes vary in terms of trip characteristics and their distribution in space and time. This leads to the heterogeneity of carbon reduction patterns associated with users' trip purposes. For example, trips connected to public transit are more likely to occur during commuting peak hours and in areas close to residential or workplace zones. Leisure trips, on the other hand, may cover longer distances than those for shopping (especially for convenience stores nearby), thus they are more likely to substitute driving or public transit, resulting in a greater environmental impact. Therefore, it is valuable to investigate the contributions of different trip purposes to the carbon reduction potential of SMSs.

To address the aforementioned research gaps, this chapter selects three Chinese cities at different stages of development as study areas to examine the potential for carbon emission reductions during the vehicle usage phase for both dockless shared bikes and e-bikes. In the estimation process, we take into account the substitution of shared micromobility for four transport modes: walking, private cars, buses, and subways. Additionally, this study investigates how trip purposes influence the carbon reduction potential of SMSs by employing a geographic information system (GIS)-based approach that calculates the spatial attractions of Points of Interest (POIs) to infer trip purposes. The findings of this study offer a comprehensive understanding of the environmental impact of SMSs across various urban contexts and different trip purposes, thereby informing relevant transport planning to achieve carbon neutrality in cities.

4.2 Literature Review

4.2.1 Carbon Reduction Estimation of Shared Micromobility

Most previous studies have examined the environmental benefits of bike-sharing, with its sustainability largely depending on its ability to replace high-emission transport modes (e.g., private cars). One of the pioneering studies, Zhang and Mi [49] utilised a big data-based approach to quantitatively estimate the environmental benefits of the dockless bike-sharing system in Shanghai. They found that shared bikes save 8,358 tonnes of fuel consumption and reduced NO_X and CO_2 emissions by 64 and 25,240 tonnes, respectively, in 2016. Li et al. [25] estimated that each trip generated by dockless bike-sharing saved 80.77 g of CO_2 emissions compared to the situation without shared bikes in Shanghai. The annual reduction in greenhouse gas emissions from shared bikes is estimated to be 117 kilotonnes of CO_2, roughly equivalent to the yearly greenhouse gas emissions of 25,000 vehicles. Diao et al. [8] developed a new approach to estimating the environmental benefits of dockless bike-sharing by combining both big and small data, taking into account both the substitutive and complementary relationships between bike-sharing and alternative transport modes. Rather than simply reporting emission reduction patterns at the regional or urban scales (e.g., [39]), Lv et al. [28] established a trip-level model to more precisely examine the carbon reduction effect of shared bikes in Shenzhen, China. They concluded that each kilometre cycled reduces 96 g of carbon emissions in the central area, with the total reduction from one day's operation amounting to approximately 114 tonnes.

In addition to the environmental impact of shared micromobility through substitution for other transport modes, the production, maintenance, recycling, and disposal processes of shared bikes and e-bikes also contribute to environmental footprints [9]. Short service life and low-frequency use of bikes are major factors that render bike-sharing less environmentally friendly [5]. Since the rebalancing of shared bikes is a significant contributor to carbon emissions, several studies have attempted to quantify the carbon emissions associated with rebalancing and incorporated this into the optimisation function during the rebalancing process [34]. However, since the focus of this chapter is to compare the environmental impacts of shared bikes and e-bikes in terms of their substitution for public transit, we only quantify the carbon reduction potential during the usage phase.

While the environmental benefits of shared bikes have been relatively well-explored, fewer studies have quantified the environmental benefits of shared e-bikes. In Zurich, Switzerland, Reck et al. [35] found that shared e-bikes emit more CO_2 than the transport modes they replace, whereas personal e-bikes result in lower CO_2 emissions compared to the modes they substitute. This challenges the widely held view in transport planning that shared micromobility systems are inherently more environmentally friendly. Focusing on Nanjing (China), Liu et al. [27] estimated the carbon emission reduction potential of a docked e-bike-sharing system. They concluded that a one-percent shift in commuter behaviour from cars to e-bikes could reduce more than 9.55 tonnes of carbon dioxide. However, their estimation was based on a hypothetical scenario where an existing docked bike-sharing system would be upgraded to e-bikes. Li et al. [22] found that the average carbon emission reduction per shared e-bike trip ranged from 108 to 120 g/km, but also emphasised that the carbon reduction potential of dockless e-bike-sharing remains uncertain. Additionally, while an increasing body of literature has compared travel patterns and relationships among different SMS types [4, 24], to the best of our knowledge, no study has quantitatively compared the carbon reduction potential of dockless e-bike-sharing with that of dockless bike-sharing.

Most previous studies have focused on a single city to investigate the environmental impact of SMSs. A few exceptions have compared cities to estimate the carbon reduction potential [19, 22]. These studies typically focused on large cities with higher levels of economic development, larger populations, and expansive administrative areas. Shanghai (China) is perhaps the most frequently studied city [7, 8, 49], though other first-tier cities, particularly in the United States, also received considerable attention [15, 35]. This focus on well-developed cities is understandable, as these cities operate most SMSs worldwide, face more significant environmental challenges, and have better data availability to support research. However, the policies, operations, and distribution of shared micromobility in medium and small-sized cities differ significantly from those in larger cities. Ignoring these "unknown" cities risks overlooking the diverse spatiotemporal patterns of micromobility use and the associated carbon emissions. This also limits the ability to provide unbiased and effective interventions for transport planning and policymaking across cities of varying sizes and developmental stages.

Finally, previous studies have used data-driven approaches to infer trip purposes for shared micromobility users. For instance, Xing et al. [45] inferred the trip purposes of shared bike users in Shanghai by combining shared bike trip data with POI information. Their findings revealed that dining is the most common trip purpose, followed by trips for "transfer to other transport modes" and shopping. Ross-Perez et al. [36] found that retail trips are more frequent than trips related to education, but cycling to residential areas is the most common purpose in Manchester. In a pioneering study, Wang and Zhang [43] compared the trip purposes of both dockless shared bikes and e-bikes in Ningbo, China. They found that dining is the most prevalent trip purpose for both SMS types, and that bike-sharing plays a more significant role in trips related to lodging and education compared to e-bike-sharing. However, contrary to the findings of Xing et al. [45], transport-related trips are found to be relatively uncommon for both bike and e-bike users. These studies have provided valuable insights into how to infer trip purposes and improved our understanding of user behaviour. Nevertheless, no study has yet explored the relationship between trip purposes and the carbon reduction potential of SMSs, leaving our understanding of trip purposes and their associated environmental benefits limited.

4.2.2 The Considered Travel Mode Substitution

Mode substitution is a critical factor in estimating the potential for carbon emission reductions. Early studies often assumed that bike trips exceeding a predefined distance (e.g., 1 km) substitute for driving trips, while those shorter than the threshold are considered to replace walking, without accounting for the potential substitution of other modes of transport, such as bus and metro services [49]. In contrast, more recent literature has sought to include the substitution effect for a wider range of travel modes to avoid both overestimating and underestimating the environmental benefits associated with SMSs. For instance, Kou et al. [19] examined the carbon emission reductions in eight major cities in the United States (e.g., Chicago, Boston, and Los Angeles) by considering the replacement of walking, cars, and different types of public transport. They found that annual carbon emission reductions ranged from 41 to 5,417 tonnes of CO_2 equivalent, with the emissions per trip varying from 283 to 581 g of CO_2 equivalent. Chen et al. [6] examined the modal shift from shared bikes to walking, buses, and cars, revealing that the docked bike-sharing system in New York City reduces 30,070 tonnes of carbon emissions between 2014 and 2017. When accounting for the substitution of dockless bike-sharing for public transit, Saltykova et al. [38] found that bike-sharing saves 13,200 L of fuel and 2,564.31 kg of CO_2 in Chengdu, China, highlighting that the environmental benefits of bike-sharing may be overestimated if the substitution for public transit is not considered. Regarding dockless e-bike-sharing, Li et al. [22] investigated how shared e-bikes replace walking, cycling, buses, and metro services, estimating that shared e-bikes reduce carbon emissions by 108–120 g per kilometre in Chengdu and Kunming, two provincial cities in southwestern China. Unlike previous studies, our research offers a comparative analysis of both dockless bike-sharing and e-bike-sharing, using three cities at different stages of development as case studies.

The most common method used to determine substitution patterns in previous empirical studies is through various mode substitution surveys. For example, Zhou et al. [51] conducted a survey to collect data on the share of different transport modes for various trip distances and their corresponding probabilities of switching to shared e-bikes in Shanghai. Respondents are typically asked retrospective counterfactual questions, such as "What transport mode would you have used for the last trip you took on a shared micromobility system if the bike/e-bike had not been available?" Response categories typically include several alternative travel modes, with an option to indicate that the trip would not have been made had the original mode not been available [35]. These surveyed substitution rates of SMSs for other transport modes are then used to estimate the environmental benefits [25]. However, these surveys are labour-intensive to conduct and often limited by sample sizes and potential biases [10, 22]. The substitution rate of a shared micromobility trip for alternative transport modes is trip-specific and can be heavily influenced by the specific urban context [25]. To overcome these limitations, more advanced methods have been proposed to identify substitution relationships using real trip data. For example, Kong et al. [17] developed a more sophisticated classification framework to identify dockless bike-sharing trips that substitute for public transit, combining trip data with travel information from online map services. This framework has proven useful in assessing the environmental impact of shared micromobility [22, 38]. This chapter will also adopt this framework and utilise massive user-generated trip data to support our analysis.

4.3 Study Area and Data

4.3.1 Study Area

This chapter selects three Chinese cities for case studies: Binjiang, Wucheng, and Xiangshan, all located in Zhejiang Province, which lies along the southeastern coast of China (Fig. 4.1). Binjiang and Wucheng are district-level cities, while Xiangshan is a county-level city. In the context of Chinese urban areas, district-level cities are typically part of larger metropolitan regions, characterised by a more developed commercial and service sector and high-tech industries. In contrast, county-level cities are smaller and often encompass surrounding rural areas, where agriculture and light industry are the primary economic activities [25]. The service area covers the entire city of Binjiang, which spans approximately 71.37 km^2 and serves a population of 530,000 residents as of 2022. Binjiang is more economically dynamic and better served by an advanced network of buses and subways compared to the other two cities. In Wucheng, the shared micromobility service area is restricted to the eastern part of the city, covering 110 km^2 (about 7.9% of the total city area) and a residential population of 489,470. Similarly, Xiangshan's SMS service area is concentrated in the eastern part of the city, covering 88.86 km^2 (roughly 5.93% of the city) and a population of around 204,110. This chapter seeks to provide a comparative analysis of the potential for SMSs to reduce carbon emissions across cities at different stages of development and varying in size. The analysis is based on the service areas where SMSs are operational.

4.3.2 Data

The data used in this chapter mainly comprises SMS trip data, POIs and street networks, explained as follows:

(1) SMS trip data. We obtain SMS trip data from Hellobike, a leading shared micromobility company in China, for the three cities. Initially, we exclude trips that are longer than 50 kms or shorter than 50 m, have a duration of over 2 h or under 1 min, or involve speeds exceeding 25 km/h. We also remove trips that fall outside the spatial boundaries of each service area. After pre-processing, the final dataset includes 314,412 shared bike trips and 205,281 shared e-bike trips, covering a one-week period from 5 to 11 September 2022. The spatial distributions of the bike trips and e-bike trips are shown in Fig. 4.2a, b. Each trip contains timestamps and coordinates of its origin and destination (OD) as well as its corresponding Trip ID, anonymous User ID, and Bike ID. The descriptive details are illustrated in Table 4.1.

(2) POIs (Fig. 4.2c). We crawl POIs via the API (Application Programming Interface) of Gaode Map, which is China's leading map service provider. We reclassify original POIs into seven categories, each of which corresponds to a specific trip purpose that will be discussed in the trip purpose analysis (Table 4.2). Particularly, the crawled public transit data, shown in Fig. 4.2d, consists of 1,240 bus stops and 112 subway stations. Using the spatial locations of public transport stations (stops), we further estimate the travel time and the number of transfers for each SMS trip, assuming public transport was used, with the help of the Baidu Map Direction API. This step is crucial to determine whether an SMS trip substitutes for a public transport trip.

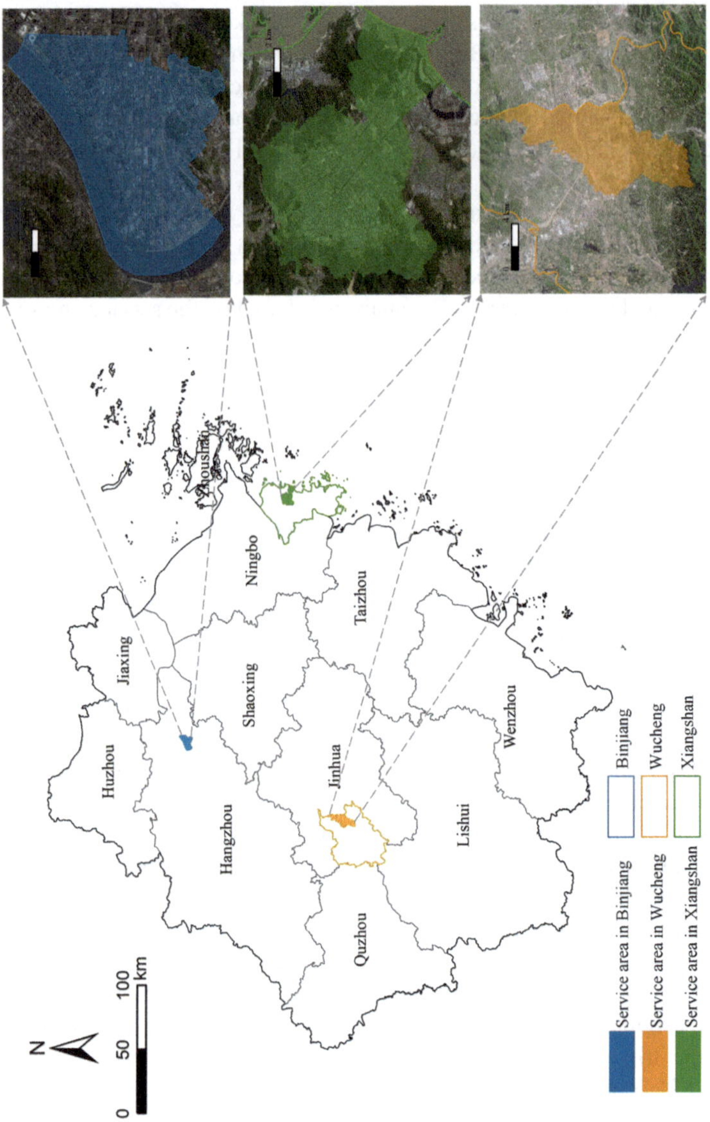

Fig. 4.1 Binjinag, Wucheng, Xiangshan and their service areas

Table 4.1 Basis descriptions of shared micromobility trips in Binjinag, Wucheng, and Xiangshan

SMS mode	City	Trip number	Bike/ e-bike number	User number	Trip duration (min)	Trip distance (km)
Bike-sharing	Binjiang	260,908	9,669	93,785	11.71	1.53
	Wucheng	24,893	1,420	6,658	11.50	0.88
	Xiangshan	28,611	2,246	9,063	13.60	1.41
	All cities	314,412	13,335	109,492	11.87	1.47
E-bike-sharing	Binjiang	71,916	1,998	37,445	17.35	2.03
	Wucheng	80,466	2,883	34,917	14.47	2.31
	Xiangshan	52,899	2,292	18,739	10.95	1.64
	All cities	205,281	7,173	91,068	14.57	2.04

(3) Street networks (Fig. 4.2e). The networks, obtained from OpenStreetMap, have a total length of 2251.69 km across the three cities, comprising 17.57 thousand street segments and 11.50 thousand street nodes.

4.4 Methodology

Figure 4.3 illustrates the analytical framework, which consists of three modules: travel mode substitution, carbon reduction calculation, and trip purpose analysis. First, this chapter establishes a model, initially developed by Kong et al. [17], to identify the potential substitution effect of SMSs based on three key factors: accessibility, duration, and convenience. The analysis considers four potential substituted travel modes: walking, driving, buses, and subways. Second, we estimate the carbon emission change for each trip by comparing the carbon emissions from the substituted modes to those from SMSs. The carbon emissions reduced from the trips are then aggregated at the grid level to better visualise and analyse the carbon emission changes caused by SMSs. Finally, to better understand the relationship between the carbon reduction potential and trip purpose, this study infers the purpose of each trip by calculating the spatial attractions of different categories of POIs.

4.4.1 Travel Mode Substitution

Following microeconomic principles, travellers assess the attributes of available options and ultimately choose the one that provides the highest subjective utility. The probability of selecting a travel mode is significantly influenced by its accessibility, along with relevant factors such as duration and convenience. This study establishes a substitution model to identify the travel mode most likely to be replaced by an SMS

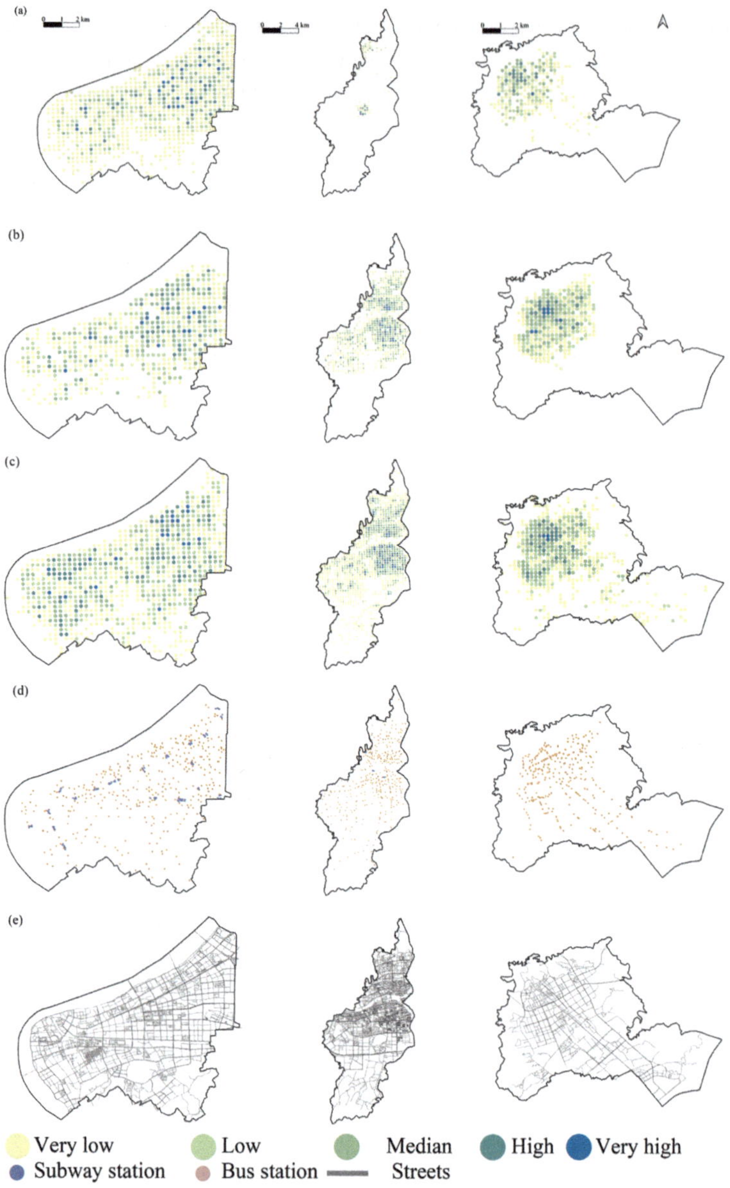

Fig. 4.2 Spatial distributions of datasets. Note: Sub-figure **a** represents shared bike trips, **b** for shared e-bike trips, **c** for POIs, **d** for public transit stations and **e** for street networks. The natural break method is used to divide the number of bike trips, e-bike trips, and POIs into the levels of very low, low, medium, high and very high

Table 4.2 POI categories and their corresponding trip purposes

POI category	Example	Corresponding purpose	POI number		
			Binjiang	Wucheng	Xiangshan
Food and beverages	Bakeries, coffee houses, restaurants, canteens and ice-cream shops	Dining	4,908	6,039	1,893
Transport facilities	bus stops and subway stations	Transfer	412 (bus) 98 (subway)	490 (bus) 14 (subway)	338 (bus) 0 (subway)
Residential places	Residential buildings and cottages	Home	6,563	5,274	1,178
Educational institutes	Schools, training institutes, research institutes and libraries	Schooling	1,848	2,245	491
Shops	Clothing stores, comprehensive markets, convenience stores, plants & pet markets and supermarkets	Shopping	5,167	11,729	4,754
Sports and recreational places	Sports stadiums, fitness centres, basketball courts, exhibition centres, art galleries, museums, planetariums, recreation centres, theatres and cinemas	Leisure	806	873	332
Working places	Companies, factories, governmental organisations, industrial parks and media organisations	Working	5,645	5,741	2,124
Total POI number			25,447	32,405	11,110

trip (Fig. 4.4). Previous studies have indicated that 500 m is a comfortable walking distance [22, 27]. Therefore, this chapter initially assumes that an SMS trip shorter than 500 m substitutes a walking trip. For SMS trips exceeding 500 m, we estimate whether they replace public transit (buses or subways),if not, we assume they substitute a driving trip. The substitution effect of an SMS trip on public transit is primarily influenced by the proximity of its origin and destination to public transit stations, its trip duration, and the number of transfers required for an alternative public transit trip [38]. The substitution relationship between SMS and public transit is determined through three steps: accessibility analysis, duration analysis, and convenience analysis.

(1) Accessibility analysis. It examines the distance from the origin and destination of an SMS trip to the nearest public transit stations. Passengers are considered

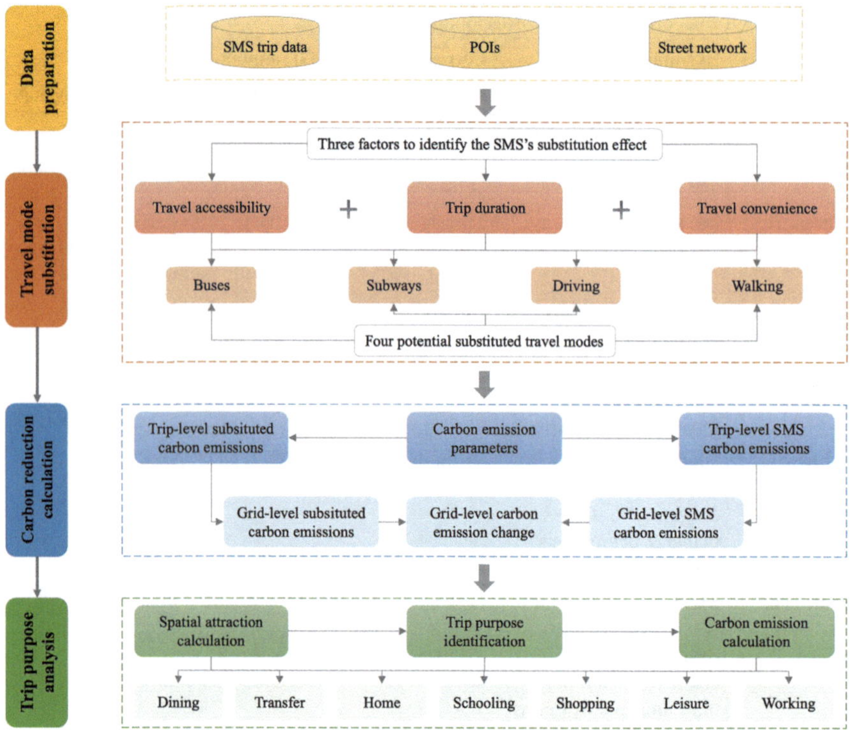

Fig. 4.3 The estimation framework of the potential of carbon emission reductions

to have access to public transit if they are within the influence area of a transit station, defined as a buffer zone around the station. We set the accessibility threshold at 400 m for bus access and 800 m for subway access [17, 22, 38]. Additionally, we exclude a 100-m radius buffer where SMS effectively integrates with public transit, meaning shared micromobility has the potential to complement rather than replace public transit. Based on these pre-determined thresholds, we divide urban areas into five zones (shown in Fig. 4.5): Zone A (within a 100-m buffer of a bus stop), Zone B (within a 100-m buffer of a subway station), Zone C (within a bus stop's buffer ranging from 100 to 400 m), Zone D (within a subway station's buffer ranging from 100 to 800 m), and Zone E (beyond these zones). We identify three scenarios based on the SMS trips' origins and destinations within these five zones. Scenario 1a and Scenario 1b include trips within the bus and subway coverage but are not close enough for convenient public transit integration. Scenario 2a and Scenario 2b demonstrate trips that start or end near bus stops and subway stations, respectively. In Scenario 3, either the trip's origin or destination is not covered by public transit, and these trips are identified as substitutes for driving rather than public transit.

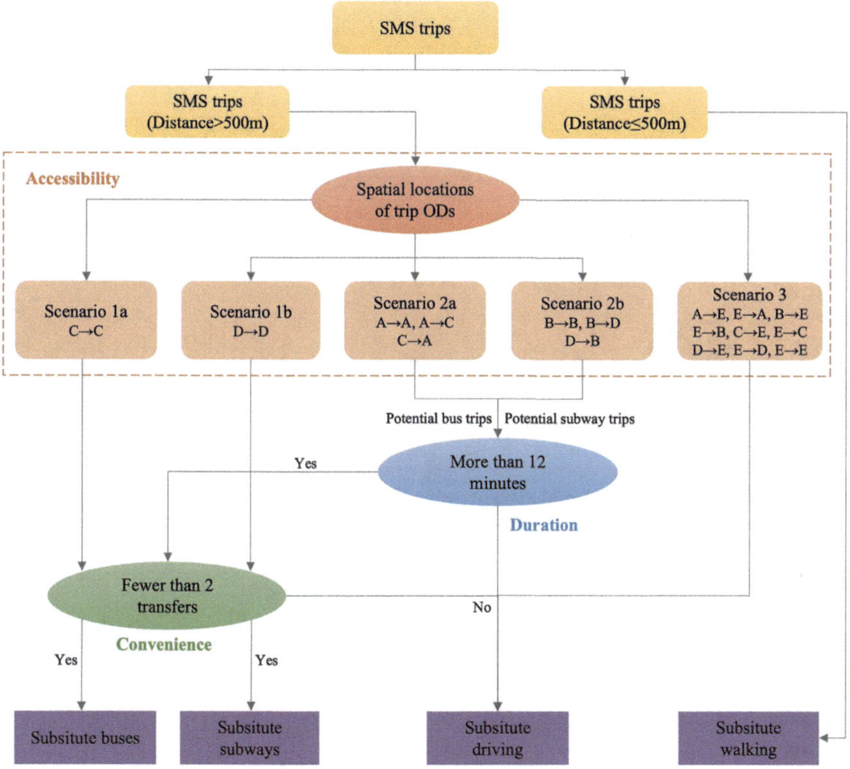

Fig. 4.4 The classification methodology of the mode substitution

(2) Duration analysis. According to Kong et al. [17], it is less likely that passengers would prefer to connect to public transit via SMS if the trip duration exceeds 12 min. Therefore, we set 12 min as the maximum tolerance time. SMS trips shorter than 12 min are classified as substitutes for driving trips, while longer trips are considered to potentially replace public transit trips based on the results of the accessibility analysis.

(3) Convenience analysis. If the conditions for accessibility and duration are met, we then consider the convenience factor. This analysis applies to trips in Scenario 1 and Scenario 2, as well as those exceeding 12 min in Scenario 3. The need for transfers may deter passengers from using public transit. This study assumes that if a trip requires more than one transfer, a traveller who would otherwise drive a car is unlikely to switch to public transit, even if it is more cost-effective. In other cases, public transit is considered the more optimal option.

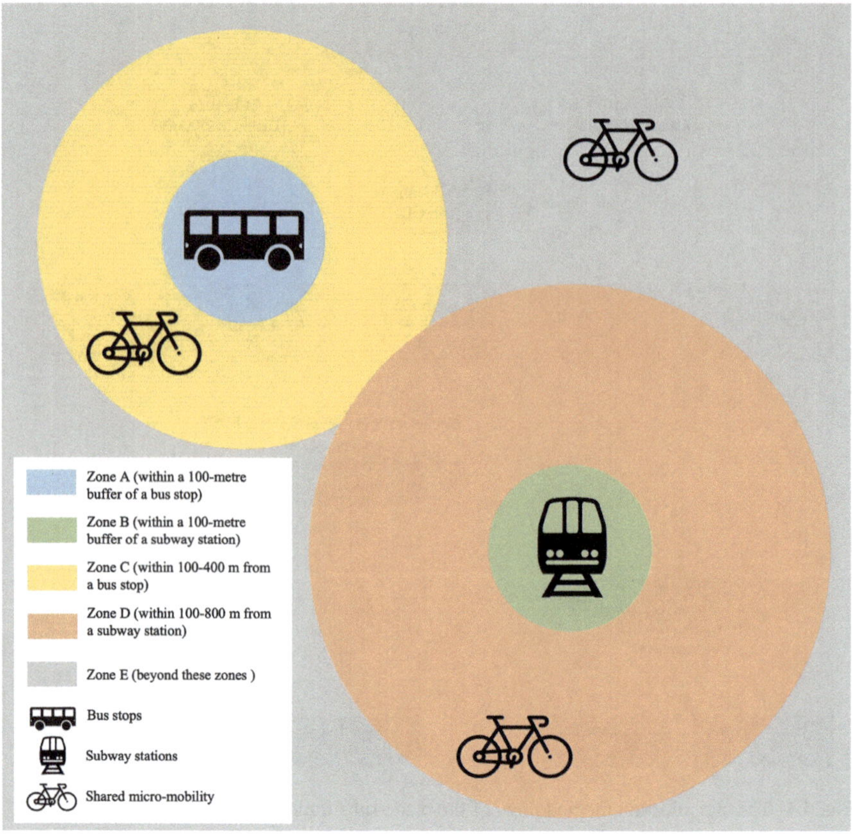

Fig. 4.5 Illustration of five zones

4.4.2 Carbon Reduction Calculation

SMSs contribute to reducing carbon emissions by shifting travellers away from carbon-intensive modes such as driving, but they may also replace trips that would otherwise be made using greener transport modes, like public transit and walking. In this sub-section, we quantify the changes in carbon emissions based on the revealed travel mode substitution effect and the carbon emission parameters of four alternative travel modes. The carbon emission parameter connects the quantity of emissions produced by a travel mode to an activity that leads to emission release. The reduction or increase in carbon emissions resulting from SMSs substituting other transport modes can be calculated by determining the difference between the carbon emissions of the substituted mode and the SMS. This is achieved by multiplying the travel distance by their respective emission parameters, shown in Table 4.3.

According to Zhang and Mi [49], the carbon emission parameter for driving a private car is set at 223.9 g/km. For buses and subways, the emission parameters

Table 4.3 Carbon emission parameters of different travel modes

Mode	Carbon emission parameter	Value (g CO_2/Pkm)	References
Driving	E_{drive}	223.90	[49]
Bus	E_{bus}	7.83	[40, 41]
Subway	E_{subway}	6.02	[40, 41]
Bike/walking	E_{bike}/E_{walk}	0.00	[49]
E-bike	E_{ebike}	10.05	[51]

that account for emissions during the usage phase are 490 g/km and 1,808 g/km, respectively [41]. In Zhejiang Province, the passenger capacity of a typical bus is approximately 60 people, while a subway can accommodate around 300 people. Consequently, the carbon emissions per passenger are estimated to be 7.83 g/km for buses and 6.02 g/km for subways [40]. Since walking and shared bikes do not require fuel consumption or produce emissions during their usage phase, their carbon emission parameter is set to 0. Shared e-bikes, however, inherently generate carbon emissions due to the electrification and charging-related activities, making them less environmentally friendly than conventional bikes [13]. The carbon emission parameter of e-bikes during the usage phase includes emissions generated by the e-bikes themselves and the trucks used for battery swapping. According to Zhou et al. [51], the CO_2 emissions generated by the direct use of shared e-bikes are 6.91 g/Pkm (passenger per kilometre), while emissions from battery swapping are 3.14 g/Pkm.

To better investigate the carbon emission reductions from SMS, we divide the city under observation into regular grids with a width and length of 250 m. Trip-level carbon emission changes are then aggregated at the grid level. Specifically, the grid-level carbon emission reduction of SMSs is formulated as:

$$E_i = \left(E_{drive} \sum_{a=1}^{A} D_a + E_{bus} \sum_{b=1}^{B} D_b + E_{subway} \sum_{c=1}^{C} D_c + E_{walk} \sum_{d=1}^{D} D_d \right) - \left(E_{bike} \sum_{m=1}^{M} D_m + E_{ebike} \sum_{n=1}^{N} D_n\right) \tag{4.1}$$

where, E_i represents SMS's total carbon emission reduction from spatial grid i. M and N are the number of shared bike and e-bike trips originating from grid i, respectively. A, B, C and D are the number of substituted driving, bus, subway and walking trips, respectively. D_a is the trip length of a substituted driving trip a, D_b is the trip length of a substituted bus trip b, D_c is the trip length of a substituted subway trip c, D_d is the trip length of a substituted walking trip d, D_m is the trip length of a shared bike trip m and D_n is the trip length of an e-bike trip n. Note that, for the city without a subway system, carbon emissions reduced from substituted metro trips are set to 0. In addition, we can calculate grid-level carbon emission reductions of shared bikes and e-bikes separately so as to investigate the environmental impact of dockless bike-sharing and e-bike-sharing from a comparative perspective. If E_i is larger than zero, shared micromobility positively impacts the environment by decreasing carbon emissions from alternative modes.

4.4.3 Trip Purpose Analysis

Urban residents use SMSs for various motivations, which influence potential substitution patterns and, in turn, affect environmental benefits [13]. Investigating the trip purposes of shared micromobility is crucial for understanding how travellers' behaviour impacts carbon emissions. Since the SMS trip data used in this study does not provide explicit information on trip purposes, we develop a GIS-based approach to infer trip purposes to overcome this limitation. Following Xing et al. [45] and Li et al. [25], we classify the trip purpose of each SMS trip into seven categories: Dining, Transfer, Home, Schooling, Shopping, Leisure, and Working, in line with the seven POI types. We then calculate the spatial attractiveness of POIs based on proximity and the proportion of surrounding POIs. Proximity is the first indicator. Unlike traditional docked systems, dockless SMSs allow users to leave bikes near their destination amenities. We use a Gaussian function to model the distance decay effect, representing the spatial attractiveness of a POI. The second factor is the proportion of POI categories within the walking radius of the destination, indicating that the more POIs within this radius, the higher the attractiveness of the POIs. The POI with the highest weighted spatial attractiveness in each category is considered the most likely travel purpose. Specifically, s_{ij} representing the normalised and weighted spatial attractiveness between locations i and j, is formulated as:

$$s_{ij} = \begin{cases} \frac{e^{-1/2(d_{ij}/d_{max})^2} - e^{-1/2}}{1-e^{-1/2}} if d_{ij} \leq d_{max} \\ 0 if d_{ij} > d_{max} \end{cases} \tag{4.2}$$

$$A(i, j, P) = s_{ij} * \frac{numberofPOIs(i, d_{max})}{totalnumberofPOIs(d_{max})} \tag{4.3}$$

$$A_{max} = \max A(i, j, P) \tag{4.4}$$

where, d_{ij} is the distance between these locations; d_{max} is the maximum radius (50 m in our case); $A(i, j, P)$ is the attractiveness of the location i for a user at location j considering POI category P; $numberofPOIs(i, d_{max})$ is the count of POIs within the radius d_{max} from location i; $totalnumberofPOIs(d_{max})$ serves as a normalising factor, representing the total count of POIs within the radius across the area.

4.5 Results

4.5.1 Travel Mode Substitution

Using the travel mode substitution model, we can assess the substitution effect of SMSs on walking, driving, buses, and subways. Among all shared bike and e-bike trips in the three cities, the majority have replaced driving, with 45.37% of shared bike trips and 45.98% of shared e-bike trips substituting for driving. In contrast, only 11.08% of shared bike trips and 7.19% of shared e-bike trips potentially replace walking. Shared e-bikes have a higher replacement ratio for subways (32.72%) than shared bikes (18.31%). Regarding trip distance, the average distances of the substituted driving, bus, and subway trips for shared bikes are 1.24 km, 2.18 km, and 1.88 km, respectively, all of which are shorter than those of shared e-bikes, with average distances of 1.61 km (driving), 2.92 km (buses), and 2.30 km (subways). The substituted walking distance for shared bikes is, on average, 0.31 km, slightly longer than that of shared e-bikes (0.27 km).

The travel mode substitution results are presented in Fig. 4.6. For shared bikes, the proportion of driving trips in the three cities ranges from 42.35 to 66.09%, while for shared e-bikes, it ranges from 34.20 to 60.50%. In terms of substituted public transit trips, the share of shared e-bikes varies from 30.11 to 59.95%, whereas for shared bikes, the range is between 13.97 and 47.58%. These figures are generally consistent with previous studies on the relationship between shared micromobility and public transit. For instance, a survey by Zhou et al. [51] found that 33.67% of shared e-bike trips substitute for public transit, while Fuller et al. [11] reported that 50.43% of shared bike users shift from public transit to shared bikes. The proportions of substituted walking trips are between 10.08 and 19.94% for shared bikes, and 5.85–9.39% for shared e-bikes. Shared bikes are more likely to replace short trips (e.g., walking trips), whereas shared e-bikes are more often used to substitute trips that would otherwise be made by private cars or public transit [51]. This is likely due to the advantage of e-bikes, which offer broader usability, alleviate physical demands, and enhance user experience over longer distances [18, 24].

In addition to the differences between shared bikes and shared e-bikes, there are significant regional differences among the three cities. The results show that the substitution of SMS trips for public transit trips is most prominent in Binjiang, whereas the substitution for driving trips is more common in Xiangshan. These differences are likely due to the considerably better public transit services in Binjiang, which has much higher densities of bus stops and subway stations. In metropolitan areas like Binjiang, shared e-bikes are more likely to replace public transit than shared bikes, which aligns with previous findings that shared e-bikes may replace trips initially made by conventional bikes and subsequently by public transit [2]. This is because shared e-bikes tend to attract users as replacements for public transit rather than integrate with it [1]. Wucheng presents a special case, as the shared bike service is limited, primarily concentrated around a university and a small northwestern part of the city, where there are only 25 accessible bus stops and no subway stations.

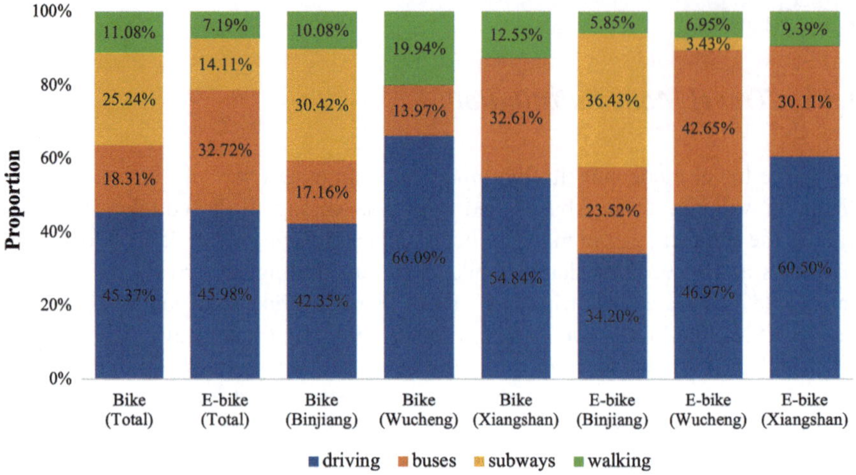

Fig. 4.6 The mode substitution effect

Due to the lack of convenient public transit service, shared bikes in this region have a higher probability of substituting walking (19.94%) and driving (66.09%). The substitution rate of shared e-bikes for driving, buses, subways, and walking in Wucheng is estimated to be 46.97%, 42.65%, 3.43%, and 6.95%, respectively. In conclusion, the variation among cities can be attributed to differences in land use, transit network densities, and SMS configurations [25].

4.5.2 *Carbon Emission Reductions*

4.5.2.1 Overall Result

Table 4.4 shows the reduced carbon emissions from shared bikes and e-bikes over the course of a week. In total, bike-sharing saves 41.51 tonnes of CO_2 across the three cities, which is higher than the potential reduction from shared e-bikes (31.84 tonnes of CO2). According to Lv et al. [28], a typical gasoline light-duty passenger vehicle emits 3.3 tonnes of carbon emissions annually, so the estimated carbon emission reduction corresponds to the emissions from 1,158 cars over a week in our study area. However, for each trip, shared e-bikes (155.11 g) reduce more carbon emissions than shared bikes (132.03 g), mainly because shared e-bike users tend to travel longer distances, thus being more likely to replace driving trips. In fact, the potential environmental benefit of shared e-bikes comes solely from the replacement of driving trips, resulting in a saving of 32.58 tonnes of CO_2. When substituting public transit and walking, e-bike-sharing has a negative environmental impact in all three cities, except for Xiangshan, which currently lacks a subway system. Similar

Table 4.4 The one-week reduced carbon emissions of shared micromobility

SMS mode and city	Sum (kg)	Average (g/trip)	Substitute for driving (kg)	Substitute for bus (kg)	Substitute for subway (kg)	Substitute for walking (kg)
Bike-sharing						
Binjiang	33,343.54	127.80	31,652.11	795.68	895.75	0.00
Wucheng	3,648.05	146.55	3,615.33	32.72	0.00	0.00
Xiangshan	4,520.03	157.98	4,365.76	154.28	0.00	0.00
Total	41,511.62	132.03	39,633.19	982.68	895.75	0.00
E-bike-sharing						
Binjiang	7,291.97	101.40	7,655.49	– 105.76	– 247.53	– 10.23
Wucheng	14,351.40	178.35	14,634.65	– 246.25	– 21.40	– 15.61
Xiangshan	10,198.22	192.79	10,295.08	– 82.96	0.00	– 13.91
Total	31,841.59	155.11	32,585.23	– 434.97	– 268.93	– 39.74

to e-bike-sharing, bike-sharing's substitution of driving trips contributes the most to the service's carbon emission reductions, particularly in Binjiang. Shared bikes contribute very little to carbon emission reductions by replacing public transit. These results clearly demonstrate that SMS is not environmentally competitive compared to buses and subways. It is worth noting that the carbon emission parameters for buses and subways in this study are based on their maximum capacity, and the actual per capita carbon emissions might be higher, especially during off-peak hours. Therefore, the real amount of carbon emissions reduced by SMS substitution for public transit is expected to be higher than the estimation shown in Table 4.4.

4.5.2.2 Spatial Pattern Analysis

To better understand the impact of the locations of SMS trips on reduced carbon emissions, we allocate the carbon emissions of each trip to its origin and aggregate them at the grid level to reveal the spatial pattern of carbon emission reductions from SMSs within a week. According to Fig. 4.7, the environmental benefit exhibits significant spatial heterogeneity. In Wucheng and Xiangshan, the reduced carbon emissions follow a radial pattern, being much higher in more central areas. This difference can be attributed to population density: the city centre is typically more populous, thus attracting more SMS vehicles and having more travellers compared to the peripheral regions [22, 49]. Additionally, the high intensities of social and economic activities in central areas contribute to higher shared micromobility ridership [25] and further amplify the environmental impact of SMSs. In Binjiang, the carbon emissions are more evenly distributed across the city, which can be largely explained by the fact that Binjiang is one of the most economically vibrant cities in Zhejiang Province (and even in China), with well-developed public transit and SMS networks covering its entire administrative area. Although in over 95% of spatial grids, the launch of

SMS contributes to reduced carbon emissions, it is still observable that shared e-bike services have led to increased carbon emission levels in certain areas (marked by the colour orange), where bus stops and subway stations are more densely distributed and are more likely to be replaced by shared e-bikes.

As shown in Fig. 4.8, the per-kilometre-based pattern differs from the distribution of overall carbon emission reductions. In the three cities (especially Binjiang and Xiangshan), shared bikes and e-bikes in non-central areas exhibit higher carbon reduction per kilometre than those in central areas. The proportions of different substituted travel modes in different cities are a key factor in explaining the spatial differences. The higher the proportion of SMS trips that replace driving trips in an area, the greater the reduction in carbon emissions and the associated environmental benefits. People living in the urban periphery, where convenient public transit services are lacking, are more reliant on private cars to access various destinations in the city centre before the introduction of SMS. Conversely, in densely populated grids, SMS is more likely to result in decreased public transit usage, as shared bikes/e-bikes offer a quicker (sometimes even faster than public transit) and more flexible alternative [29]. This finding is also recognised at the city level. Compared with the district-level city of Binjiang, Xiangshan is a county-level city consisting of large rural areas with fewer public transit services. For shared bikes, the average carbon emission reduction is 98.00 g/km in Binjiang and 125.34 g/km in Xiangshan, while for shared e-bikes, it is 70.56 g/km and 127.77 g/km, respectively. These factors contribute to variations in the substituted travel modes, affecting the carbon emission reductions at both grid and city levels.

4.5.2.3 Temporal Pattern Analysis

From a temporal perspective, we also analyse the carbon emission reductions of shared bikes and shared e-bikes over a week (Fig. 4.9). Overall, evident differences are observed between the temporal distributions of weekdays (10.90 tonnes of CO_2 per day) and weekends (9.43 tonnes of CO_2 per day). However, the trip-level carbon emission reductions across weekends (141.88 g) and workdays (140.89 g) show little difference. Therefore, the carbon reduction differences across days may be mainly attributed to day-level variations in ridership. When comparing shared bikes and e-bikes, they show similar temporal patterns of carbon emission changes during weekdays and weekends. Binjiang's reduced carbon emissions from shared e-bikes are much lower than those of shared bikes at almost every hour throughout the week, but the carbon emission changes in the remaining two cities show an opposite trend. This is simply because Binjiang benefits the most from bike-sharing, while Wucheng and Xiangshan are better served by shared e-bikes.

The temporal carbon emission changes in shared bikes and e-bikes follow typical daily commuting patterns. Overall, there are noticeable reductions during morning peaks (7–8 am) and evening peaks (5–8 pm), which save 37.18 tonnes of carbon emissions, accounting for 50.69% of the total reduction within a day. Specifically, reduced carbon emissions during the evening peaks are higher than during the morning ones in

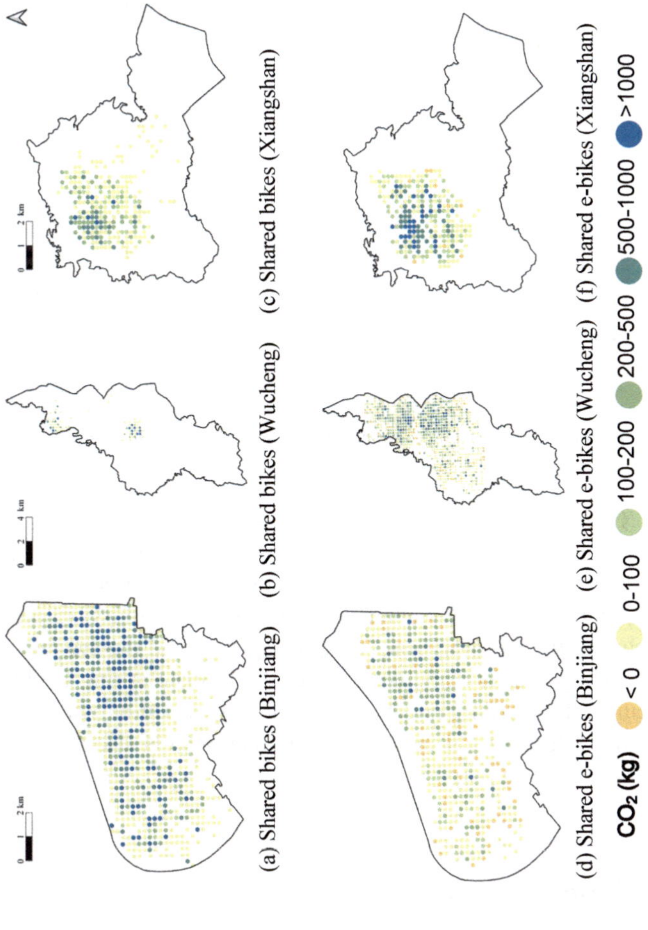

Fig. 4.7 Spatial distributions of total carbon emission reductions

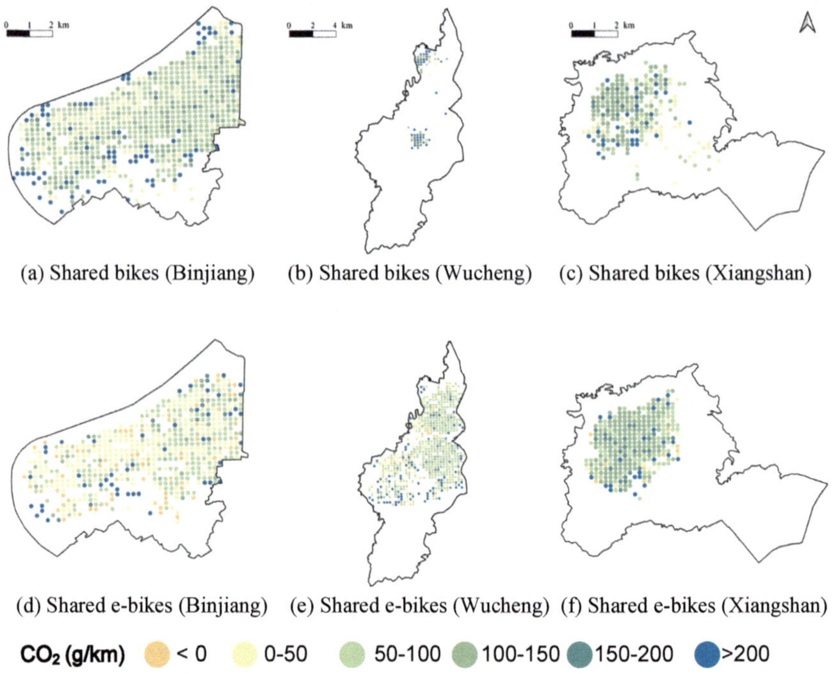

Fig. 4.8 Spatial distributions of per-kilometre based carbon emission reductions

Wucheng and Xiangshan, implying more diverse and flexible activities for cyclists after work [6]. This distribution differs from car-sharing, which typically experiences higher morning peaks [46]. In the case of Binjiang, a district-level city with a much larger population and more vibrant economy, its carbon emissions are reduced slightly more during the morning peaks of working days and the evening peaks of weekends, suggesting that its commuting patterns dominate during working days. For all three cities, the morning and evening peaks of carbon emission reductions are less pronounced during weekends, indicating that SMS is also a popular mode for various non-commuting purposes [32].

4.5.3 Trip Purpose Analysis

Table 4.5 shows the inferred percentages of trip purposes. For shared bikes, the main trip purposes are Dining (40.56%), Shopping (18.07%), Working (16.89%), and Home activities (9.36%). Shared e-bikes are most frequently used for Shopping (37.98%), followed by Dining (33.01%) and Working activities (15.68%). The inferred results are similar to the findings from Xing et al. [45]: catering and shopping are more popular purposes of bike-sharing. They are also consistent with our

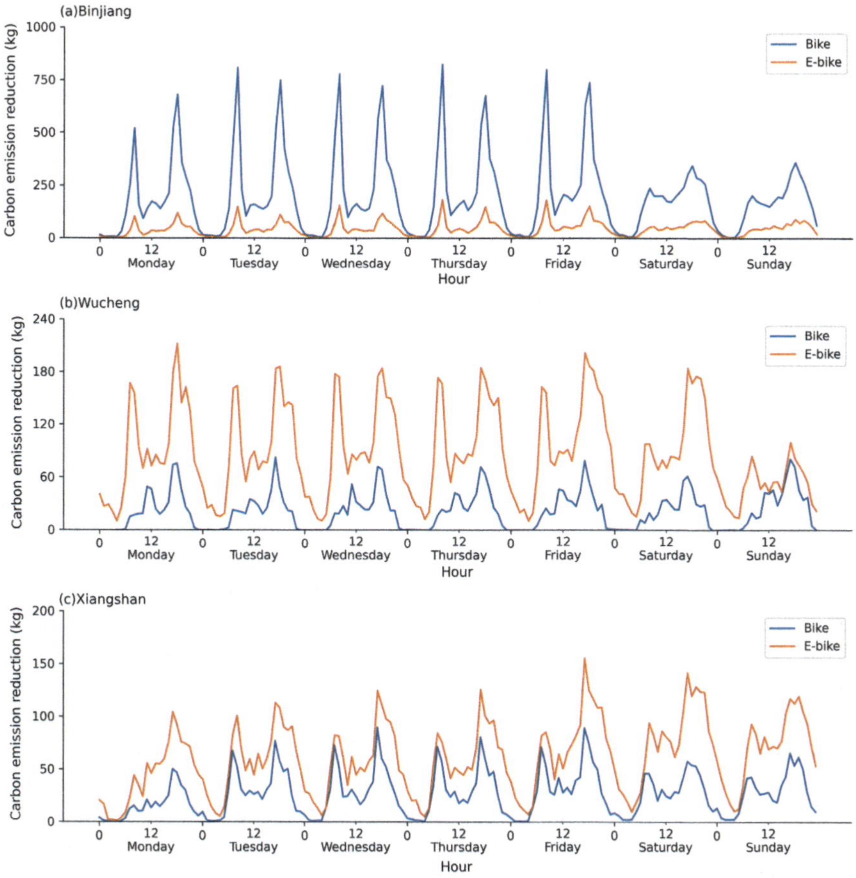

Fig. 4.9 Temporal distributions of carbon emission reductions

empirical knowledge that SMS is an attractive mode for daily commuting. Transfer is not a significant trip purpose, as we expected, since connecting to public transit is often claimed as a role of SMSs by service providers and transport planners. The proportion of shared bikes is 9.70%, and of shared e-bikes is 5.06%, accounting for 11.59% and 4.99% of carbon emission reductions. This proportion is lower than that estimated by Xing et al. [45], who found that about 19.58% of shared bike trips are Transfer-related in Shanghai, but similar to the findings from Wang and Zhang [43], who inferred that shared bike and e-bike trips with the purpose of transport account for 7.8% and 7.0% of all trips in Ningbo – a city also located in Zhejiang Province, China. Two reasons could explain this relatively low proportion of Transfer purposes. First, people living in a megacity like Shanghai tend to rely more on public transit systems, while those living in smaller cities, like those discussed in this study, may have more commuting options, resulting in fewer Transfer-related trips [43]. Second, despite using POIs to infer trip purposes being widely accepted, it could be biased

Table 4.5 Percentages of trips by trip purpose

Trip purpose	Total		Binjiang		Wucheng		Xiangshan	
	Bike (%)	E-bike (%)	Bike (%)	E-bike (%)	Bike (%)	E-bike (%)	Bike (%)	E-bike (%)
Dining	40.56	33.01	40.03	39.00	49.34	31.89	37.73	26.58
Transfer	9.70	5.06	11.53	8.93	0.06	2.46	1.39	3.78
Home	9.36	4.08	8.61	5.35	22.42	4.06	4.80	2.39
Schooling	4.30	3.14	3.39	4.04	16.40	3.47	2.05	1.40
Shopping	18.07	37.98	16.56	22.05	6.96	42.35	41.45	53.00
Leisure	1.12	1.05	0.99	0.98	3.17	1.40	0.51	0.60
Working	16.89	15.68	18.87	19.65	1.65	14.38	12.06	12.26

when inferring the Transfer purpose, because the number of transport-related POIs would be much lower than other POI categories, making it less represented in our calculation process.

Table 4.6 shows the percentages of reduced carbon emissions by trip purpose. Generally, purpose-based carbon emission reductions are linearly associated with trip percentages by purpose. Trips with Dining and Shopping purposes account for about 57.43% and 72.10% of total reduced carbon emissions in the three cities for shared bikes and e-bikes, respectively. As a more economically vibrant city with a well-developed public transit network and SMS, Binjiang reduces more carbon emissions from Transfer and Working trips. The Leisure trips contribute the least to carbon emission reductions in all cities, but Wucheng's Leisure trips have a greater reduction contribution than Binjiang and Xiangshan. In addition, for shared bikes, Wucheng has much higher proportions of carbon emissions reduced from Home and Schooling purposes, even higher than Working and Shopping. This is reasonable mainly because the bike-sharing service covered a university campus there. Shopping trips of shared e-bikes are dominant in reducing carbon emissions in Xiangshan, accounting for 53.09% of total emissions and implying a good SMS service in the city's commercial area.

4.6 Conclusions and Discussion

This chapter estimated and compared the potential for carbon emission reductions from shared micromobility systems (SMSs) in three Chinese cities (i.e., Binjiang, Wucheng, and Xiangshan) using real user-generated trip data. It also investigated the environmental benefit contributions of trips with different purposes. In terms of SMS's travel mode substitution effect, the results showed that most shared bike and e-bike trips potentially substitute driving, accounting for 45.37% and 45.98% of total trips, respectively. However, shared bikes are more likely to replace walking trips, whereas shared e-bikes are more often used to substitute public transit trips.

Table 4.6 Percentages of carbon emission reductions by trip purpose

Trip purpose	Total		Binjiang		Wucheng		Xiangshan	
	Bike (%)	E-bike (%)	Bike (%)	E-bike (%)	Bike (%)	E-bike (%)	Bike (%)	E-bike (%)
Dining	40.76	31.20	39.98	39.84	47.57	29.89	41.03	26.87
Transfer	11.59	4.99	14.20	11.03	0.07	2.68	1.66	3.93
Home	9.58	4.18	8.71	5.73	23.46	4.56	4.81	2.52
Schooling	4.01	2.73	3.03	3.39	16.03	3.34	1.53	1.39
Shopping	16.67	40.90	14.81	20.17	7.62	42.77	37.70	53.09
Leisure	1.10	1.31	0.90	0.90	3.73	1.98	0.40	0.66
Working	16.29	14.70	18.37	18.93	1.51	14.79	12.87	11.54

For SMS's environmental benefit, shared bikes reduce emissions by 41.51 tonnes of CO_2, which is higher than the 31.84 tonnes of CO_2 reduced by shared e-bikes over a week in the three cities. In contrast, at the trip level, a shared bike potentially reduces 132.03 g of CO_2, which is lower than a shared e-bike's 155.11 g. Most of the reduced carbon emissions are from driving substitution, and shared e-bikes even generate a negative environmental impact when substituting public transit and walking. From a spatial perspective, carbon emission reductions are concentrated in the central areas of Wucheng and Xiangshan, whereas Binjiang's reduction in carbon emissions is more widely distributed across the whole city. When it comes to the per-kilometre-based pattern of carbon emission reductions, SMSs reduce more carbon emissions in the peripheries of the three cities, mainly because more driving trips are replaced in these non-central areas. Temporally, SMS's carbon reduction patterns are similar to typical temporal usage patterns: there are more carbon emissions reduced during morning/evening commuting peaks and on weekdays. Regarding trip purposes, the results show that SMS trips with Dining and Shopping purposes contribute the most to carbon emission reductions, whereas Leisure trips have the least environmental benefit. Compared to Wucheng and Xiangshan, Binjiang has more carbon emissions reduced from Transfer and Working trips. Shared bike trips with Home and Schooling purposes reduce more carbon emissions in Wucheng.

Our findings provide valuable insights for urban authorities and enterprises to optimise SMSs to promote car substitution, reduce the replacement of public transit and walking, and ultimately improve urban sustainability. First, both shared bikes and e-bikes are encouraged to be adopted in cities with smaller sizes and less developed stages, because SMSs in these cities (e.g., Wucheng and Xiangshan) are more likely to substitute private car usage and serve as a good alternative to public transit, which is relatively underdeveloped compared to that in Binjiang. Second, many cities empirically encourage the implementation of shared micromobility as a first/last-mile connection to public transit, but this strategy should be approached with caution, considering the other potential role of SMSs. The presented study shows that the substitution effect of SMSs on public transit can be significant: 18.31% and 25.24% of shared bike trips have replaced buses and subways, respectively, whereas the proportions of shared e-bike trips substituting buses and subways reach 32.72% and 14.11%, respectively. Therefore, maintaining a good balance between encouraging SMS-public transit integration and avoiding SMS substitution of public transit is essential to improve the overall performance of urban transport systems. Third, while allocating more shared bikes and e-bikes in central areas is reasonable to increase environmental benefits and trip usage, it is also important to plan more cycling infrastructure and distribute more shared vehicles in the urban peripheries, as these areas perform better in terms of per-kilometre carbon emissions. In addition, developing SMSs in the suburban areas of a city is associated with social equity benefits, as many disadvantaged populations, such as rural migrants, may reside there [50]. Finally, better rebalancing and battery-swapping strategies are required to ensure that users can conveniently use SMSs during commuting peak times on weekdays when SMSs generate the most reduced carbon emissions. Similar strategies are necessary to ensure the convenient use of SMSs in areas close to catering, shopping, and working places.

Our study also has several limitations. Firstly, new energy vehicles, which are becoming increasingly popular in China and globally, use electricity either partially or entirely as an energy source, resulting in much lower carbon emissions compared to traditional cars [28]. As of December 2022, the penetration rate of new energy vehicles in Zhejiang Province stood at 35.8%, which exceeds the national average by 13.9 percentage points (Tide [42]). Clearly, it is essential to account for the trend of transport electrification, particularly in our study areas, all of which are located in Zhejiang Province. Furthermore, this chapter only considers changes in carbon emissions during the usage phase, excluding emissions generated during the manufacturing, recycling, and rebalancing processes, which may lead to an overestimation of the environmental benefits of SMSs. Future research could improve upon this by adopting a life-cycle assessment approach. Lastly, this study infers trip purposes by calculating the spatial attractiveness of POIs surrounding the destination of each SMS trip. However, this proportion-based method may be biased, particularly for POIs whose categories are less represented in the dataset but which could be significant in defining trip purposes, such as transport facilities. In future studies, we will aim to infer trip purposes by combining POI importance with their proportions, supported by more advanced inference techniques such as Latent Dirichlet Allocation [43].

References

1. Bieliński T, Kwapisz A, Ważna A (2021) Electric bike-sharing services mode substitution for driving, public transit, and cycling. Transp Res Part Transp Environ 96:102883
2. Bigazzi A, Wong K (2020) Electric bicycle mode substitution for driving, public transit, conventional cycling, and walking. Transp Res Part Transp Environ 85:102412
3. Bretones A, Marquet O (2023) Riding to health: investigating the relationship between micromobility use and objective physical activity in Barcelona adults. J Transp Health 29:101588

4. Campbell AA, Cherry CR, Ryerson MS, Yang X (2016) Factors influencing the choice of shared bicycles and shared electric bikes in Beijing. Transp Res Part C Emerg Technol 67:399–414
5. Chen Z, Van Lierop D, Ettema D (2020) Dockless bike-sharing systems: what are the implications? Transp Rev 40:333–353. https://doi.org/10.1080/01441647.2019.1710306
6. Chen Y, Zhang Y, Coffman D, Mi Z (2022) An environmental benefit analysis of bike sharing in New York City. Cities 121:103475
7. Cheng B, Li J, Su H, Lu K, Chen H, Huang J (2022) Life cycle assessment of greenhouse gas emission reduction through bike-sharing for sustainable cities. Sustain Energy Technol Assess 53:102789
8. Diao M, Song K, Shi S, Zhu Y, Liu B (2023) The environmental benefits of dockless bike sharing systems for commuting trips. Transp Res Part Transp Environ 124:103959
9. D'Almeida L, Rye T, Pomponi F (2021) Emissions assessment of bike sharing schemes: the case of Just Eat Cycles in Edinburgh. UK. Sustain Cities Soc 71:103012
10. Fukushige T, Fitch DT, Handy S (2021) Factors influencing dock-less E-bike-share mode substitution: evidence from Sacramento, California. Transp Res Part Transp Environ 99:102990
11. Fuller D, Gauvin L, Kestens Y, Morency P, Drouin L (2013) The potential modal shift and health benefits of implementing a public bicycle share program in Montreal, Canada. Int J Behav Nutr Phys Act 10:66. https://doi.org/10.1186/1479-5868-10-66
12. Fyhri A, Fearnley N (2015) Effects of e-bikes on bicycle use and mode share. Transp Res Part Transp Environ 36:45–52
13. De Haas M, Kroesen M, Chorus C, Hoogendoorn-Lanser S, Hoogendoorn S (2022) E-bike user groups and substitution effects: evidence from longitudinal travel data in the Netherlands. Transportation 49:815–840. https://doi.org/10.1007/s11116-021-10195-3
14. He Q, Musterd S, Boterman W (2022) Understanding different levels of segregation in urban China: a comparative study among 21 cities in Guangdong province. Urban Geogr 43:1036–1061
15. He P, Zou Z, Zhang Y, Baiocchi G (2020) Boosting the eco-friendly sharing economy: the effect of gasoline prices on bikeshare ridership in three US metropolises. Environ Res Lett 15:114021
16. Hu Y, Zhang Y, Lamb D, Zhang M, Jia P (2019) Examining and optimizing the BCycle bike-sharing system—a pilot study in Colorado. US. Appl Energy 247:1–12
17. Kong H, Jin ST, Sui DZ (2020) Deciphering the relationship between bikesharing and public transit: modal substitution, integration, and complementation. Transp Res Part Transp Environ 85:102392
18. Kontar W, Ahn S, Hicks A (2022) Electric bicycles sharing: opportunities and environmental impacts. Environ Res Infrastruct Sustain 2:035006. https://doi.org/10.1088/2634-4505/ac7c8b
19. Kou Z, Wang X, Chiu SFA, Cai H (2020) Quantifying greenhouse gas emissions reduction from bike share systems: a model considering real-world trips and transportation mode choice patterns. Resour Conserv Recycl 153:104534
20. Lamb WF, Wiedmann T, Pongratz J, Andrew R, Crippa M, Olivier JG et al (2021) A review of trends and drivers of greenhouse gas emissions by sector from 1990 to 2018. Environ Res Lett 16:073005
21. Li X, Cottam A, Wu Y-J, Khani A (2020) Can a bikesharing system reduce fuel consumption? Case study in Tucson, Arizona. Transp Res Part Transp Environ 89:102604
22. Li Q, Fuerst F, Luca D (2023) Do shared E-bikes reduce urban carbon emissions? J Transp Geogr 112:103697
23. Li A, Gao K, Zhao P, Qu X, Axhausen KW (2021) High-resolution assessment of environmental benefits of dockless bike-sharing systems based on transaction data. J Clean Prod 296:126423
24. Li Q, Zhang E, Luca D, Fuerst F (2024) The travel pattern difference in dockless micromobility: shared e-bikes versus shared bikes. Transp Res Part Transp Environ 130:104179
25. Li S, Zhuang C, Tan Z, Gao F, Lai Z, Wu Z (2021) Inferring the trip purposes and uncovering spatio-temporal activity patterns from dockless shared bike dataset in Shenzhen. China. J Transp Geogr 91:102974

26. Lin D, Zhang Y, Meng L (2023) Assessing bike accessibility to metro systems by integrating crowdedness. Trans Urban Data Sci Technol 27541231231179403
27. Liu Y, Tian Z, Pan B, Zhang W, Liu Y, Tian L (2022) A hybrid big-data-based and tolerance-based method to estimate environmental benefits of electric bike sharing. Appl Energy 315:118974
28. Lv G, Zheng S, Chen H (2024) Spatiotemporal assessment of carbon emission reduction by shared bikes in Shenzhen, China. Sustain Cities Soc 100:105011
29. Van Marsbergen A, Ton D, Nijënstein S, Annema JA, Van Oort N (2022) Exploring the role of bicycle sharing programs in relation to urban transit. Case Stud Transp Policy 10:529–538. https://doi.org/10.1016/j.cstp.2022.01.013
30. McKinsey. Shared mobility: sustainable cities, shared destinies 2023. https://www.mckinsey.com/industries/automotive-and-assembly/our-insights/shared-mobility-sustainable-cities-shared-destinies. Accessed 26 March 2024
31. Nikitas A (2018) Understanding bike-sharing acceptability and expected usage patterns in the context of a small city novel to the concept: a story of 'Greek Drama.' Transp Res Part F Traffic Psychol Behav 56:306–321
32. Nosal T, Miranda-Moreno LF (2014) The effect of weather on the use of North American bicycle facilities: a multi-city analysis using automatic counts. Transp Res Part Policy Pract 66:213–225. https://doi.org/10.1016/j.tra.2014.04.012
33. Orvin MM, Fatmi MR (2021) Why individuals choose dockless bike sharing services? Travel Behav Soc 22:199–206
34. Qin M, Wang J, Chen W-M, Wang K (2023) Reducing CO2 emissions from the rebalancing operation of the bike-sharing system in Beijing. Front Eng Manag 10:262–284. https://doi.org/10.1007/s42524-021-0168-y
35. Reck DJ, Martin H, Axhausen KW (2022) Mode choice, substitution patterns and environmental impacts of shared and personal micromobility. Transp Res Part Transp Environ 102:103134
36. Ross-Perez A, Walton N, Pinto N (2022) Identifying trip purpose from a dockless bike-sharing system in Manchester. J Transp Geogr 99:103293
37. Rérat P (2021) The rise of the e-bike: towards an extension of the practice of cycling? Mobilities 16:423–439. https://doi.org/10.1080/17450101.2021.1897236
38. Saltykova K, Ma X, Yao L, Kong H (2022) Environmental impact assessment of bike-sharing considering the modal shift from public transit. Transp Res Part Transp Environ 105:103238
39. Shang W-L, Chen J, Bi H, Sui Y, Chen Y, Yu H (2021) Impacts of COVID-19 pandemic on user behaviors and environmental benefits of bike sharing: a big-data analysis. Appl Energy 285:116429
40. Shao D, Li H (2021) Carbon reduction benefits and carbon peak goals of urban passenger transportation electrification: a case study of Shanghai. Urban Transp China 19:53–58+42
41. Tian P, Zhang H, Mao B, Zhang S (2024) Comparison of carbon emission intensity of different urban passenger transport modes. China Environ Sci 44:2823–2832
42. Tide News (2023) The "88 Strategy" 20th anniversary of the practice of people's livelihoods 2023. https://www.zj.gov.cn/art/2023/12/5/art_1229740710_60185355.html. Accessed 28 June 2024
43. Wang C, Zhang Y (2023) Inferring and comparing trip purposes of shared micromobility services: a case study of Ningbo, China. Trans Urban Data Sci Technol 27541231231179884
44. World Meteorological Organization (2015) WMO Greenhouse Gas Bulletin
45. Xing Y, Wang K, Lu JJ (2020) Exploring travel patterns and trip purposes of dockless bike-sharing by analyzing massive bike-sharing data in Shanghai, China. J Transp Geogr 87:102787. https://doi.org/10.1016/j.jtrangeo.2020.102787
46. Yu B, Ma Y, Xue M, Tang B, Wang B, Yan J et al (2017) Environmental benefits from ridesharing: a case of Beijing. Appl Energy 191:141–152. https://doi.org/10.1016/j.apenergy.2017.01.052
47. Zhang Y, Kong H, Lin D (2023) Editorial: Shared micromobility and future cities. Trans Urban Data Sci Technol 2:155–158. https://doi.org/10.1177/27541231231202575

48. Zhang Y, Lin D, Mi Z (2019) Electric fence planning for dockless bike-sharing services. J Clean Prod 206:383–393
49. Zhang Y, Mi Z (2018) Environmental benefits of bike sharing: a big data-based analysis. Appl Energy 220:296–301
50. Zhang Y, Song Y, Zhang W, Wang X (2024) Working and residential segregation of migrants in Longgang City, China: a mobile phone data-based analysis. Cities 144:104625
51. Zhou Y, Yu Y, Wang Y, He B, Yang L (2023) Mode substitution and carbon emission impacts of electric bike sharing systems. Sustain Cities Soc 89:104312

Chapter 5
Analyse Spatial Accessibility and Equity of Dockless Bike-Sharing

5.1 Introduction

Bike-sharing is a form of transport that caters to residents' daily needs for short-distance, low-speed travel through rental services. Thanks to its convenience and sustainability, bike-sharing has been widely promoted and adopted globally in recent decades, with China leading in both scale and growth of service usage [10]. According to a report by the China Road Transport Association, as of March 2023, the total number of shared bikes deployed in China exceeded 14.9 million, with approximately 27 million daily trips on average. In the same year, the bike-sharing market in North America reached a value of $415 million [48]. Current bike-sharing systems can be broadly categorised into docked and dockless forms. Unlike docked systems, dockless bike-sharing does not require fixed parking stations. Users can easily locate and use bikes via smartphones, park them at designated spots (i.e., electronic fences), and securely lock them, completing the electronic payment process once they reach their destination [58]. Bike-sharing systems have expanded urban residents' transport options, playing a crucial role in solving the 'first mile' and 'last mile' commuting challenges, while also helping to reduce energy emissions. They are key in supporting the development of a multi-layered and diversified sustainable urban transport system [52, 59].

The rapid expansion of bike-sharing services has sparked significant interest and research on the issue of service equity. As a crucial component of urban transport systems, the development of equitable bike-sharing services offers considerable benefits for enhancing the sustainability and inclusivity of urban mobility [52]. A substantial body of research has addressed this topic using diverse data sources, including surveys, real bike-sharing trip data, and demographic statistics. These studies have examined the demographic profiles of bike-sharing users and analysed the spatial distribution of shared bike accessibility, highlighting disparities in service utilisation across various urban regions and demographic groups [20, 25, 43].

Y. Zhang, *Understanding Shared Micromobility in China Using Massive User-Generated Trip Data*, Urban Sustainability,
https://doi.org/10.1007/978-981-96-8492-2_5

However, existing research predominantly centres on Western cities and station-based bike-sharing systems, with limited analysis of Chinese cities and dockless bike-sharing systems (DBSs). China, with its vast scale of DBS deployment and large user base, presents a unique context for examining service equity. Given the marked differences in demographic characteristics and socio-economic environments between Chinese and Western cities, studying DBS in Chinese urban settings can reveal equity patterns that diverge from those observed in the West. Furthermore, current methods for measuring the accessibility of DBS fail to fully account for both the accessibility of bike-sharing services and the opportunities available within urban areas. Few studies incorporate the attractiveness of various opportunities or activities into accessibility calculations, yet such factors are essential for capturing the travel preferences and usage patterns of bike-sharing services, ultimately enhancing the accuracy and comprehensiveness of accessibility assessments.

To address this gap, this chapter uses Ningbo, a city in Zhejiang Province, China, as a case study. It introduces an enhanced measure of accessibility, termed Service-Opportunity Accessibility (SOA), which integrates both the accessibility of dockless bike-sharing services via walking and the accessibility of urban opportunities via cycling. This combined approach is used to evaluate the service levels of DBS across different areas of the city. By applying the SOA measure, the study assesses the equity of DBS from both horizontal and vertical perspectives, examining how the distribution of SOA varies across different urban regions and socio-economic groups. The analysis draws on multiple data sources and provides valuable insights for the planning and management of DBS, aiming to enhance service equity and optimise urban transport strategies.

5.2 Literature Review

5.2.1 Bike-Sharing Analysis

The existing literature on bike-sharing has primarily focused on, but is not limited to, topics such as examining system characteristics and their influencing factors, analysing external impacts, and optimising system design and planning.

The first topic focuses on uncovering the usage patterns of bike-sharing services and identifying the factors that influence these patterns. Numerous studies have examined spatiotemporal variability of bike-sharing usage and the purposes behind trips. For example, research conducted in Singapore has shown that, on weekdays, bike-sharing trips are predominantly concentrated during the morning and evening peak periods, with trip origins and destinations mostly situated near residential areas, central business districts, and rapid transit stations [60]. Studies on dockless bike-sharing have found that trips on weekends tend to be longer than on weekdays, with primary trip purposes including dining, shopping, transit transfers, and commuting [55, 56]. Usage patterns are also influenced by personal preferences, personality traits, and socio-cultural factors [17, 51]. Additionally, factors such as the built environment (including land use and road connectivity), travel time, and weather conditions significantly impact bike-sharing demand and travel characteristics [15]. For instance, Li et al. [33] applied a geographically weighted regression model to explore the factors affecting bike-sharing utilisation. Their findings revealed a positive correlation between bike-sharing usage and factors such as the public transport service coverage, population density, traffic near metro stations, and the diversity of points of interest (POIs). Furthermore, Ji et al. [26] investigated the relationship between bike-sharing usage and street configurations as well as the built environment in Xi'an (China). They discovered that enhancing the direct connectivity of street networks can increase bike-sharing usage, while also identifying an inverted U-shaped relationship between land use intensity and bike-sharing utilisation.

Topic two centres on evaluating the societal, economic, health, and environmental benefits associated with bike-sharing, as well as its impact on other modes of transport, such as buses and subways. Existing studies suggest that the integration of bike-sharing with public transit tends to be more synergistic than substitutive or complementary [30]. By seamlessly integrating with public transport and substituting for other transport modes, bike-sharing can help alleviate traffic congestion, reduce greenhouse gas emissions, and improve public health outcomes, thereby providing substantial economic and social advantages. Building on this, some researchers have attempted to predict the transport modes displaced by each bike-sharing trip, based on factors such as trip duration, origin, and destination, in order to quantify the environmental benefits of bike-sharing in terms of reduced fuel consumption and emissions. This approach also offers valuable insights for the planning and management of urban bike-sharing systems [31, 50, 59]. Furthermore, Clockston and Rojas-Rueda [9] conducted a comprehensive analysis of the health impacts of physical activity, air pollution, and traffic accidents. Their study found that bike-sharing systems in the United States could prevent approximately 4.7 premature deaths annually and generate an estimated $36 million in health-related economic benefits.

Topic three focuses on the optimisation and planning of bike-sharing systems, addressing key aspects such as station rebalancing, demand forecasting, and bike-sharing station siting. For example, Fu et al. [16] developed a two-stage robust model to optimise bike-sharing station locations and rebalancing operations, with the aim of maximising operational revenue by improving bike-sharing utilisation rates. Similarly, Chen et al. [6] proposed an advanced distribution robust optimisation method for optimising dockless bike-sharing rebalancing. This approach integrates the roles of workers and trucks in rebalancing operations and effectively addresses uncertainties in bike-sharing demand. Furthermore, some studies leveraged built environment and accessibility factors to predict bike-sharing service demand across various travel purposes and scenarios. This allows for the identification of bike-sharing station locations that are better aligned with demand and accessibility, thereby enhancing the fairness and accessibility of bike-sharing services [5, 13, 40].

The studies mentioned above integrate bike-sharing services into the broader urban system, offering a systematic analysis of the characteristics of bike-sharing services and their interactions with urban environments and components. They highlighted the pivotal role of bike-sharing in achieving urban sustainability, providing valuable insights for the understanding and planning of bike-sharing services within cities.

5.2.2 Bike-Sharing's Accessibility Analysis

In transport planning, accessibility refers to the potential to engage with opportunities or facilities [22]. It is commonly used to reflect the ease with which residents can access opportunities and services, or the extent of spatial barriers in specific areas [37]. As a key indicator of the transport services available in a given area, accessibility is closely linked to whether urban residents can reach adequate public services or employment opportunities, thereby significantly influencing their quality of life and urban inclusivity [1]. Consequently, accessibility plays a crucial role in formulating and evaluating transport policies, as well as in planning and optimising public transport and service facilities [18]. Various methods have been employed to measure accessibility, such as the cumulative opportunity measure [37], the gravity-based model [19], and the two-step floating catchment area (2SFCA) method [21]. These approaches have also been applied to assess bike-sharing accessibility, helping to analyse the equity of this service and inform its planning practices [5, 11].

Previous studies on bike-sharing accessibility can be broadly categorised into two main types. One type focuses on accessibility to bike-sharing facilities and services. These studies typically use the locations of bike-sharing stations as a basis, assessing the quantity and convenience of services that residents in a particular location can access within a defined spatial and temporal range, often by calculating distances to facilities or the coverage level of service areas. For example, Wang and Lindsey [54] evaluated bike-sharing accessibility based on the walking distance from users to bike-sharing stations. They found that reduced accessibility has a significantly negative impact on the frequency of bike-sharing usage. Jin and Sui [27] constructed bike-sharing catchment areas using a 400-m Euclidean distance boundary, identifying whether research units overlap with service areas to assess accessibility and discover inequities in access across different demographic groups. Taking into account both demand and service capacity, Chen et al. [5] employed the two-step floating catchment area (2SFCA) method to measure the spatial accessibility of bike-sharing stations in Nanjing, China. They found that establishing new stations in areas with low accessibility greatly expands the service area of bike-sharing. This type of research tends to view bike-sharing solely as a service facility, rather than as a mode of transport. The measured accessibility reflects the ease of individual access to bike-sharing services (i.e., bike-sharing service accessibility), but it does not capture the quantity or potential of other services or facilities that users can access through bike-sharing (i.e., bike-sharing opportunity accessibility). The latter better aligns with the concept and significance of accessibility and is the focus of the second type of research.

The second type of research on bike-sharing accessibility incorporates users' potential destinations and accessible amenities while measuring access to bike-sharing facilities, thus reflecting users' opportunity accessibility through the use of bike-sharing services. For instance, when investigating the dockless bike-sharing system in Shanghai, Li et al. [32] introduced trip purposes into the measurement of bike-sharing accessibility and employed the gravity-based model to assess and compare accessibility under different trip purposes. Their findings revealed that bike-sharing accessibility is lower for commuting to work or school. Similarly, Qian and Jaller [47] calculated accessibility by considering the number of various POIs accessible within a certain range and their importance in bike-sharing trips, demonstrating that improving accessibility positively influences the likelihood of disadvantaged groups using bike-sharing. Furthermore, Giuffrida et al. [20] developed a comprehensive indicator to measure bike-sharing accessibility by integrating the service level of bike-sharing facilities, the convenience of accessing POIs, and their attractiveness for bike-sharing usage. This indicator was used to reveal differences in bike-sharing accessibility across various areas of Dublin (Ireland). This type of research takes into account the quantity, ease of access, and attractiveness of the opportunities and amenities users can reach via bike-sharing. It offers a more accurate reflection of users' purposes in utilising bike-sharing and highlights the role of bike-sharing services as a tool for urban residents to access opportunities and resources.

However, previous studies often base their assessment of the attractiveness of opportunities on limited sample surveys [47] or assign equal weight to different facilities [20], making it challenging to accurately reflect the differences in attractiveness among various facilities for bike-sharing trips. These differences in attractiveness can reveal residents' actual travel preferences, which in turn influence the prediction of trip purposes and the calculation of opportunity accessibility. For example, if users in a specific area primarily use bike-sharing to connect to public transport, bike-sharing facilities that provide access to a greater number of public transport stations within a certain riding range tend to have higher opportunity accessibility, as users can access more of the opportunities they seek through these facilities. Furthermore, although there is an increasing focus on investigating the accessibility of dockless bike-sharing systems, the existing literature on bike-sharing accessibility predominantly centres on dock-based systems, with limited and insufficient discussion on DBS. This gap in the literature makes it difficult to assess accessibility in DBS, as it cannot be based on fixed docking stations.

To address these gaps, this study proposes an enhanced model called Service-Opportunity Accessibility to measure the accessibility of DBS by integrating both service accessibility for residents and opportunity accessibility for DBS trips. In this model, we directly use the coordinates of bike-sharing trip parking points and opportunities to assess the attractiveness of various opportunities (i.e., POIs). Additionally, a kernel density function-based balanced floating catchment area (KDBFCA) method is employed to measure the accessibility of DBS, considering both the availability of shared bikes and their opportunity accessibility.

5.2.3 Bike-Sharing's Equity Analysis

Equity can be understood as the fair distribution of benefits or costs among different groups [8]. It is a prominent research topic in the field of urban transport [12, 46, 61]. The analysis of transport equity is typically approached from two perspectives: horizontal equity and vertical equity. Horizontal equity focuses on ensuring equal status or weight for all groups in the distribution of benefits or costs, aiming to provide equal access to resources and services for all populations. However, this perspective tends to overlook the differences in resource and service needs among various groups. In contrast, vertical equity stresses the need for resource allocation to favour disadvantaged groups, thereby reducing the cost burden on them and enhancing the overall fairness of the system [3, 35, 57].

Existing studies have shown that bike-sharing services offer positive benefits, such as improving urban transport accessibility, enhancing the city environment, and promoting resident health [41, 52]. However, the distribution of bike-sharing services and their associated benefits is not equitable. From the perspective of horizontal equity, some scholars have calculated bike-sharing accessibility in different urban areas and used methods like Lorenz curves and Gini coefficients to assess the fairness of the allocation of bike-sharing services. The results indicated that the distribution of bike-sharing services across urban spaces and populations is unequal. Bike-sharing facilities tend to concentrate in city centres or areas with higher socioeconomic status, leading to fewer facilities and reduced accessibility in urban suburbs and areas with higher levels of deprivation, despite residents in these areas often relying more heavily on bike-sharing as a low-cost and convenient transport option [12, 42, 44]. Regarding vertical equity, socio-economic attributes such as gender, age, income, occupation, race, and education level significantly influence the equity of bike-sharing usage. Bike-sharing users are more likely to be young, male, and have higher income and education levels [25, 53]. Studies focusing on North America have found that higher-income groups and Caucasians are more likely to be included in bike-sharing service areas, while communities with higher poverty levels, and concentrations of women, minorities, and older adults, have limited access to bike-sharing facilities [27].

The research discussed provides a comprehensive analysis of equity issues in bike-sharing systems. However, existing studies on bike-sharing equity primarily relied on survey and demographic data to examine the distributional differences in bike-sharing accessibility among populations with varying socioeconomic characteristics. There is a notable gap in research that utilises new urban data sources, such as mobile phone data, which can more accurately reflect the spatial distribution and socioeconomic attributes of populations. Moreover, existing studies are predominantly focused on cities in North America and Europe, with limited attention given to cases in Chinese cities. China has a large-scale bike-sharing system and user base, and the socio-economic characteristics of its residents differ significantly from those in Western countries. Therefore, studying the differences in bike-sharing service accessibility among populations with varying socio-economic attributes in Chinese cities is both necessary and valuable. To address this gap, this study focuses on Ningbo, Zhejiang

Province, as a case study in China. It primarily utilises dockless bike-sharing trip data, mobile phone data, and POIs to analyse the differences in SOA of DBS across urban spaces and population groups. This analysis aims to provide insights for optimising dockless bike-sharing services in Chinese cities.

5.3 Study Area and Data

5.3.1 Study Area

Ningbo, a city in Zhejiang Province, is selected as the case study for this research. Located on the southeast coast of China, it is one of the country's fifteen sub-provincial cities and comprises ten administrative districts (Fig. 5.1a). Due to the limited coverage of DBS, which is primarily concentrated in densely populated and economically active urban centres, this study focuses on the central urban area of Ningbo. This area includes parts of the districts of Zhenhai, Jiangbei, Haishu, Yinzhou, and Beilun (Fig. 5.1b).

5.3.2 Data

The core data used in this study is provided by HelloBike, Ningbo's largest DBS provider. The dataset includes 1,692,082 records of dockless shared bike trips from September 6 to 19, 2021. Each trip record contains information such as trip ID, bike ID, user ID, trip duration, trip distance, and the latitude and longitude of both the starting and parking points. Prior to analysis, abnormal trips with durations of less than 1 min or more than 120 min, and trips with cycling distances of less than 50 m, are excluded. These anomalous records, which are typically caused by bicycle malfunctions or users forgetting to lock their bikes, are unlikely to accurately reflect users' travel purposes and characteristics. After cleaning the data, 1,637,709 valid trip records are retained. Most trips occur on weekdays, particularly during the morning and evening peak hours (Fig. 5.2).

The POI data is crawled from Gaode Map, comprising a total of 752,988 entries, initially classified into 23 categories, such as transport and shopping facilities. Each record includes the POI's name, geographical coordinates, and category. The original data categories are adjusted by removing duplicate and irrelevant POIs. Categories related to automobiles and motorcycles are merged into a single category, and hotels, apartments, and other accommodation facilities are combined into a new category named "residential facilities". After preprocessing, 14 categories are retained, with a total of 702,914 POI entries (Table 5.1). Figure 5.1c presents the distribution heatmap of POIs. It is evident that areas closer to the centre of the study area exhibit higher POI density, suggesting more concentrated economic activities in these regions.

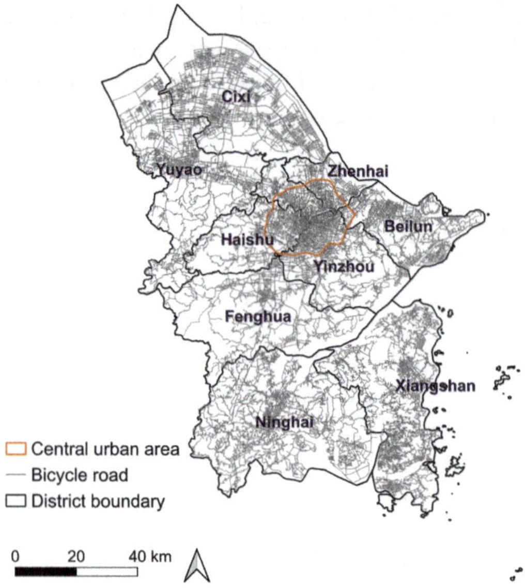

(a) The city of Ningbo

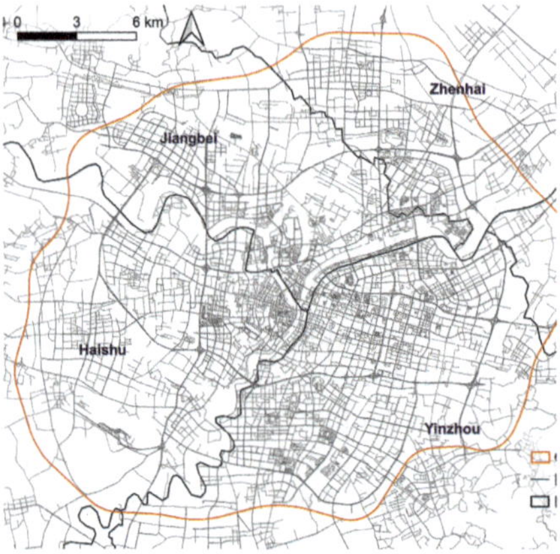

(b) Ningbo's main city zone

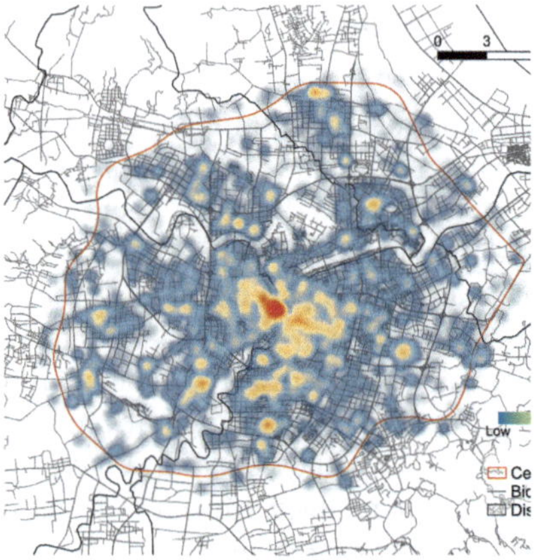

(c) The POI heatmap

Fig. 5.1 The study area and datasets

China Mobile, one of China's three major mobile communication operators, provides the mobile phone data used in this study. The dataset includes information on the spatial coordinates of users' residential locations, identified through visit records captured by communication base stations at specific times, as well as social attributes such as gender, age, and province of household registration (*hukou*). This

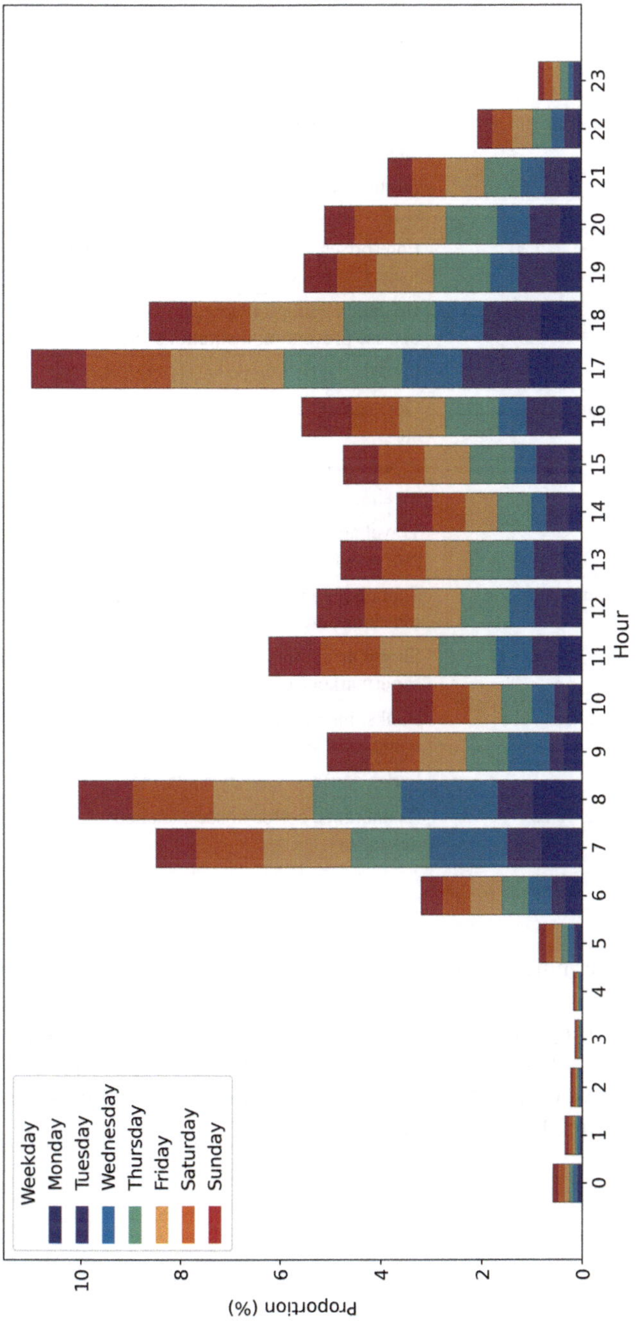

Fig. 5.2 Temporal distribution of shared bike trips

Table 5.1 Description of POIs

POI category	POI examples	Number
Transport services	Airports, ports and marinas, ropeway stations, subway stations, bus stops, railway stations, parking lots, etc	30,640
Sports and recreational facilities	Sports stadiums, recreation centres, theatres and cinemas, etc	9,327
Public facilities	Public phones, public toilets, emergency shelters, etc	8,180
Companies	Enterprises, companies, animal husbandries, fishery bases, etc	88,707
Health care facilities	Hospitals, clinics, emergency centres, disease prevention institutions, etc	13,926
Residential facilities	Residential buildings, hotels, etc	183,521
Government and social organisations	Government agencies, social organisations, etc	22,956
Car related services	Filling stations, car washes, car sales, car repairs, etc	16,934
Daily life services	Travel agencies, post offices, telecom offices, professional service firms, job centres, etc	77,904
Science and education institutions	Schools, scientific research institutions, museums, etc	20,398
Shops	Shopping centres, convenience stores, supermarkets, etc	149,992
Finance and insurance services	Banks, insurance companies, finance companies, etc	7,608
Tourist attractions	Parks, squares, scenery spots, etc	4,417
Food and beverage services	Restaurants, coffee houses, tea houses, dessert houses, etc	69,482
Total		702,914

data is used to determine the spatial distribution of populations and their associated social attributes.

Furthermore, the cycling road network, obtained from OpenStreetMap using the OSMnx tool, is a topological network consisting of 144,885 edges and 58,472 nodes (Fig. 5.1). This network will serve as a fundamental element for the network analysis conducted using ArcGIS 10.8 software.

5.4 Methodology

This study develops an SOA model to investigate the spatial equity issues of DBS in Ningbo City. The model integrates both service accessibility and opportunity accessibility of DBS. Service accessibility reflects the ease of accessing DBS, while opportunity accessibility refers to access to various urban opportunities (i.e., POIs) via DBS.

The methodology of this study consists of three main steps (Fig. 5.3). First, the locations and available shared bikes within geofences are identified using the Hierarchical Density-Based Spatial Clustering of Applications with Noise (HDBSCAN) algorithm. The opportunity accessibility of these geofences is then calculated based on their locations and surrounding POIs. Second, the spatial distribution of residents and their demographic attributes in the study area are determined using mobile phone data. The kernel function-based balanced floating catchment area (KDBFCA) method is employed to calculate the SOA of BAZs in different areas, taking into account the population size of BAZs and the service capacity of geofences. Finally, the equity performance of the DBS is assessed from both horizontal and vertical equity perspectives.

The following sub-sections provide a detailed explanation of the specific methods used in this study.

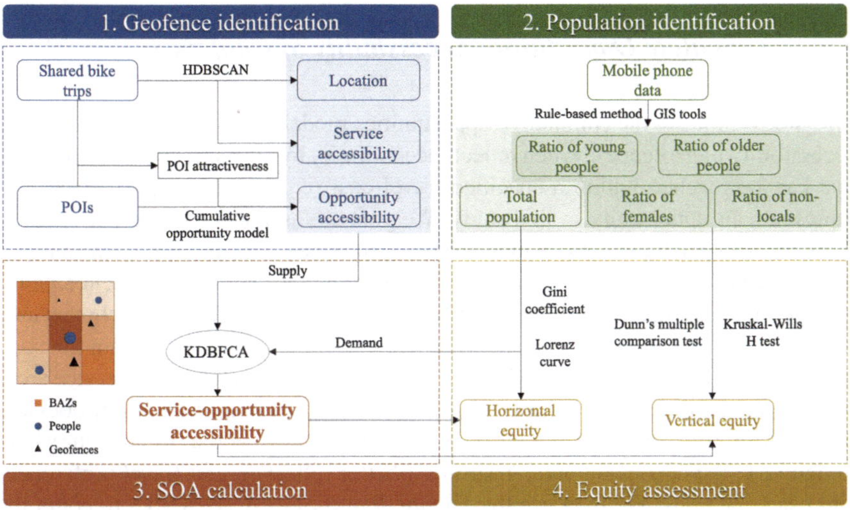

Fig. 5.3 The analytical framework

5.4.1 Identify Locations and Availability of Geofences

The calculation of service accessibility is based on the parking locations of shared bikes. However, dockless bike-sharing systems do not have fixed parking stations, which increases computational complexity when directly using the parking locations of individual trips. Therefore, this study first extracts the coordinates of parking points from DBS trips and identifies parking hotspots as geofences by employing the HDBSCAN algorithm. HDBSCAN is a density-based clustering algorithm that calculates the local density of each point, connects these points into a hierarchical clustering tree, and then identifies clusters of different densities by cutting the tree [2]. It does not require a predefined number of clusters and performs better with fewer parameter dependencies compared to the DBSCAN algorithm, which has been frequently used to identify geofences in DBS [58]. In this study, the parking points of bike-sharing trips are used as sample points, with the minimum cluster size parameter set to 14. This setting ensures that each geofence has at least one bike-sharing trip record per day over a two-week period (i.e., September 6–19, 2021), controlling cluster size and excluding abnormally small clusters. The centroids of these clusters are taken as the coordinates of the geofences, and the cluster size divided by 14 represents the number of daily available bikes for each geofence, indicating its availability.

5.4.2 Calculate Opportunity Accessibility of Geofences

This study employs the cumulative opportunities model to calculate the opportunity accessibility of geofences, which reflects accessibility by the number of opportunities that can be reached within a specified travel cost (e.g., distance) [37]. Each POI represents an opportunity in this study. Weights are assigned to different types of POIs according to their attractiveness to shared bike users when summing the number of accessible POIs. To achieve this, we first generate the cycling catchment area for each geofence using its coordinates and the cycling road network. The catchment area represents the spatial extent that users can reach by cycling within a certain time and distance from the geofence. Based on previous studies investigating the service areas of transport facilities [14, 36], the 75th percentile of cycling distances of dockless bike-sharing trips starting from each geofence is selected as the distance threshold to delineate the catchment area, ensuring it better reflects users' travel preferences and usage patterns. We then use the *Spatial Join* tool in ArcGIS 10.8 to count the number of POIs within each catchment area, followed by a weighted aggregation of these counts to calculate the opportunity accessibility of the geofences. Specifically, the opportunity accessibility of geofence k, namely O_k, can be expressed as:

$$O_k = \sum_{j=1}^{J} a_j \mathrm{N}_{kj} \tag{5.1}$$

where, N_{kj} represents the number of POIs of type j within the catchment area of geofence k, and a_j denotes the attractiveness weight for POIs of type j.

Existing studies have employed various methods to measure the attractiveness of opportunities and facilities when inferring the purposes and patterns of dockless bike-sharing trips. For example, Xing et al. [56] used the K-means + + clustering algorithm to identify the primary purposes of dockless bike-sharing trips in Shanghai. They estimated the spatial attractiveness of POIs based on the density of different types of POIs within the maximum walking radius from the origin and destination points, as well as their nearest distances to these points. Similarly, Wang and Zhang [55] used the Latent Dirichlet Allocation (LDA) model to infer the trip purposes of dockless bike-sharing trips and extracted fourteen structured travel patterns from regular bike-sharing trips. Ross-Perez et al. [49] further incorporated factors such as time, service capacity, and daily activity dynamics to calculate the attractiveness of POIs in Shenzhen and Manchester. These studies share a common assumption: DBS users tend to park bikes as close to their destinations as possible. In line with this assumption, this study uses the K-Dimensional Tree (KD-Tree) algorithm to identify the nearest POI to the parking location as the target POI for a specific trip. The KD-Tree is a common algorithm that utilises the partition trees technique for neighbour queries. It constructs a K-dimensional tree spatial index for K-dimensional space points that need to be matched, allowing for rapid retrieval of the nearest neighbour points within the tree-structured K-dimensional subspace. The KD-Tree algorithm demonstrates significantly lower computational complexity and excellent performance in neighbour queries in low-dimensional spaces compared to directly calculating and comparing distances between points [7]. The attractiveness of a particular POI type can be quantified by the proportion of dockless bike-sharing trips it draws. It can be expressed as:

$$a_j = \frac{NT_j}{NT} \tag{5.2}$$

where, NT represents the total number of dockless bike-sharing trips, NT_j is the number of trips with the POIs of type j as the target, and a_j should satisfy $\sum_{j=1}^{J} a_j = 1$.

5.4.3 Aggregate Social Attributes of Residents

By integrating the residential locations and social attributes of users from mobile phone data, we can reveal the spatial distribution of residents and their social characteristics within the study area. This is crucial for calculating accessibility and conducting equity analysis. The process can be divided into two steps:

Grid Creation: The study area is divided into 500 m x 500 m grids using the *Create Fishnet* tool in ArcGIS 10.8. Each grid is assigned a unique ID, and these grids serve as the basic analysis zones (BAZs) in this study.

Spatial Matching and Attribute Calculation: The mobile phone data is converted into a point feature layer based on the residential coordinates of users. We then use the *Spatial Join* tool in ArcGIS 10.8 to match these point features with the grids, allowing the aggregation of population attributes for each BAZ. Specifically, we calculate the following aggregated attributes for each grid: population counts by age group, the number of males and females, and the number of local and non-local residents.

5.4.4 Calculate SOA for BAZs

The SOA of each BAZ consists of two components: service accessibility and opportunity accessibility. Service accessibility is quantified by the average number of available shared bikes that residents within a BAZ can access. Opportunity accessibility reflects the number of preferred POIs residents can reach within a certain cycling range from geofences that are accessible from the BAZ. Drawing on the identified geofence locations, their bike availability, and opportunity accessibility, we calculate the SOA of BAZs using the KDBFCA method. This approach allows us to integrate both the supply of shared bikes and the demand for opportunities, providing a nuanced measure of DBS accessibility.

The Floating Catchment Area (FCA) method is a well-established approach for measuring accessibility to public facilities such as urban parks and healthcare centres [21, 29]. This method accounts for distance decay—the idea that the influence of a facility decreases with distance—and adjusts for the distribution of both services and demand by using flexible catchment areas rather than fixed boundaries. The calculation process of the FCA method mainly consists of two steps:

Step one involves calculating the service demand of each facility based on their spatial weights with corresponding demand points:

$$P_j = \sum_{i=1}^{n} P_i w_{ij} \tag{5.3}$$

where, the service demand of facility j, denoted as P_j is determined by weighting the demand quantity (e.g. population counts) of demand points P_i located within its catchment area, which is defined as the maximum range within which residents can access services. w_{ij} signifies the spatial relationship weight between demand point i and facility j, typically derived from a distance decay function, which implies that facilities closer to demand points will be allocated a higher demand quantity since individuals prefer accessing services closer to their location. Then, the service level of the facility j is calculated as:

$$L_j = \frac{S_j}{P_j} = \frac{S_j}{\sum_{i=1}^{n} P_i w_{ij}} \tag{5.4}$$

where, S_j represents the service capacity of the facility j.

The second step is to compute the accessibility of demand point i by weighting the service levels of facilities located within its catchment area based on their spatial weights w_{ij}, which can be expressed as:

$$A_i = \sum_{j=1}^{J} L_j w_{ij} = \sum_{j=1}^{J} \frac{S_j w_{ij}}{\sum_{i=1}^{n} P_i w_{ij}} \tag{5.5}$$

While traditional FCA captures the competitive dynamics of demand points in accessing services, it may overestimate demand and supply due to overlapping catchment areas [11]. To address this concern, Paez et al. [45] introduced an enhancement to traditional floating catchment area methods, known as the Balanced Floating Catchment Area method (BFCA), which normalises the spatial weights between demand points and facilities. Using w_{ij}^i represents the adjusted spatial weight between demand point i and the facilities within its catchment area and J denotes the number of facilities within the catchment area. This process can be shown as:

$$w_{ij}^i = \frac{w_{ij}}{\sum_{j=1}^{J} w_{ij}} \tag{5.6}$$

It will ensure that the services allocated based on spatial weights match the actual service capacity of the facilities, and the same applies to demand allocation as shown below:

$$w_{ij}^j = \frac{w_{ij}}{\sum_{i=1}^{n} w_{ij}} \tag{5.7}$$

The adjusted spatial weights satisfy the following conditions:

$$\sum_{j=1}^{J} w_{ij}^i = 1 \tag{5.8}$$

$$\sum_{i=1}^{n} w_{ij}^j = 1 \tag{5.9}$$

This study employs the geofences of DBS as facilities and the centroids of each BAZ as the locations of demand points. The service capacity of each station is represented by the average daily availability of bikes and their opportunity accessibility, while the population of each BAZ is used to reflect the demand quantity. The SOA of BAZ i can be expressed as:

$$A_i = \sum_{j=1}^{J} \frac{S_j w_{ij}^j}{\sum_{i=1}^{n} P_i w_{ij}^i} = \sum_{j=1}^{J} \frac{u_j w_{ij}^j}{\sum_{i=1}^{n} P_i w_{ij}^i} O_j \tag{5.10}$$

where, u_j represents the availability of the geofence j and O_j denotes the opportunity accessibility of these shared bikes. The spatial weights between geofences and demand points are calculated using the quadratic kernel function, based on walking distance. This function assigns lower weights to target points near the distance boundaries and decreases weights more sharply as distance increases, making it well-suited for spatial weight calculations in DBS. The formulation of the quadratic kernel function is presented as follows:

$$w_{ij} = \frac{3}{4}\left[1 - \left(\frac{d_{ij}}{d_0}\right)^2\right] \tag{5.11}$$

where, d_{ij} represents the walking distance between facility j and demand point i, and d_0 ($d_{ij} \leq d_0$) represents the maximum walking distance acceptable for users to access DBS, which is also the distance threshold for the catchment area of facilities and demand points. Previous studies often set this value between 300 and 1,000 m [23, 28, 39].

This study adopts distance thresholds of 500 m, 750 m, and 1,000 m. It compares the SOA results of BAZs and the equity analysis across these thresholds to ensure the robustness of the conclusions. The OD cost matrix is generated in ArcGIS 10.8 to produce a walking distance matrix between geofences and demand points, facilitating the identification of geofences and demand points within the catchment area.

5.4.5 Analyse Equity Performance

By integrating population data with the SOA for each BAZ, the equity analysis of DBS is conducted from both horizontal and vertical equity perspectives.

For horizontal equity, the study evaluates the fairness of accessibility distribution among the population within the study area. This is achieved using the Gini index and Lorenz curve, both of which are effective tools for assessing the equality of benefit distribution across the population. The Gini index can be formulated as:

$$G = 1 - \sum_{k=1}^{n}(X_k - X_{k-1})(Y_k + Y_{k-1}) \tag{5.12}$$

Here, X_k denotes the cumulative population proportion of the first to the kth BAZ, relative to the total population, while Y_k represents the cumulative SOA values of the first to the kth BAZ, relative to the total SOA values. The Gini coefficient ranges between 0 and 1, with a higher value indicating greater benefit distribution inequality. Additionally, the study visualises the spatial distribution characteristics of SOA across BAZs.

Vertical equity focuses on the distribution of benefits and costs among disadvantaged groups. Previous studies have commonly used indicators such as age, gender, and income to identify levels of disadvantage. Women, the elderly, and young people often face barriers in accessing bike-sharing services and benefiting from cycling [27, 43, 53], and are therefore a key focus of this study. Additionally, this study includes non-local populations in the analysis — individuals residing in Ningbo without local household registration (*hukou*). This group primarily consists of the floating population, who migrate from rural areas in less economically developed provinces of China to work in more prosperous cities, often in labour-intensive sectors. Without local household registration, they are excluded from public welfare schemes such as medical insurance and work injury compensation, placing them at a relative economic and social disadvantage [24, 34, 38].

The vertical equity of DBS is evaluated by examining the relationship between SOA and the proportions of females, youth (individuals under 18 years old), elderly (those aged 65 and above), and non-locals across BAZs using the Kruskal–Wallis H test (KW test). The KW test is a non-parametric method based on ranks, suitable for comparing whether there are significant differences in the distribution of a continuous or ordinal variable across more than two independent samples. It does not require the assumption of normally distributed samples, making it broadly applicable [4]. To conduct the analysis, BAZs are categorised into five levels using the Natural Breaks Classification method based on their accessibility. The proportions of youth, elderly, females, and non-locals are then tested for significant differences across these groups. Following the KW test, Dunn's multiple comparison test is applied to identify which pairs of groups exhibit significant differences. This test is also non-parametric and rank-based, making it robust to non-normal distributions. To further enhance the reliability of the results, the Holm stepwise adjustment is employed to correct for multiple comparisons, reducing the risk of type I errors (i.e., falsely rejecting a true null hypothesis).

5.5 Results

5.5.1 Identified Geofences, Availability and Opportunity Accessibility

The location and daily availability of shared bikes at geofences are determined using the HDBSCAN algorithm. A total of 7,916 geofences are identified within the study area, with each geofence providing an average of 15 shared bikes per day. Additionally, the KD-Tree algorithm is employed to assess the relative attraction of different types of POIs to DBS trips. This is based on the spatial coordinates of parking locations for DBS trips and POIs, which aids in calculating the opportunity accessibility for each geofence. The POI attraction results, as shown in Table 5.2, reveal that residential facilities and transport services are the most attractive to shared bike users, followed by shopping and food and beverage services. This indicates that DBS in Ningbo is primarily utilised for commuting home, linking to transit stations, and

Table 5.2 The POI attractiveness

POI category	Attractiveness
Transport services	0.18
Sports and recreational facilities	0.02
Public facilities	0.01
Companies	0.05
Health care facilities	0.02
Residential facilities	0.23
Government and social organisations	0.03
Car related services	0.02
Daily life services	0.10
Science and education institutions	0.07
Shops	0.14
Finance and insurance services	0.02
Tourist attractions	0.01
Food and beverage services	0.11
Total	1

accessing shopping and dining services. Opportunity accessibility for each geofence is calculated by summing the number of various types of POIs covered within its cycling catchment area, using the relative attractiveness of each POI type as weight coefficients. To reduce data variability and minimise the impact of outliers, a log transformation is applied to the calculated values of availability and opportunity accessibility for each geofence.

Figure 5.4 illustrates the spatial distribution of geofences, along with their availability and opportunity accessibility. It is evident that the spatial distribution of availability appears to be random, while opportunity accessibility shows a decreasing trend from the centre to the periphery of the study area. The spatial distribution of opportunity accessibility is primarily influenced by built environment factors, including the density of bike lanes and POIs. As depicted in Fig. 5.1, the central area of the study zone has a more concentrated bike road network and a higher density of POIs, which results in more accessible opportunities for cycling within a given distance. This contributes to the higher opportunity accessibility observed in geofences located in the central area.

5.5.2 Calculated Population Attributes of BAZs

The population attributes of the BAZs are determined using mobile phone data and GIS tools. BAZs without any population presence are excluded from the analysis, as they have minimal impact on accessibility and equity assessments. Figure 5.5

(a) The availability of geofences

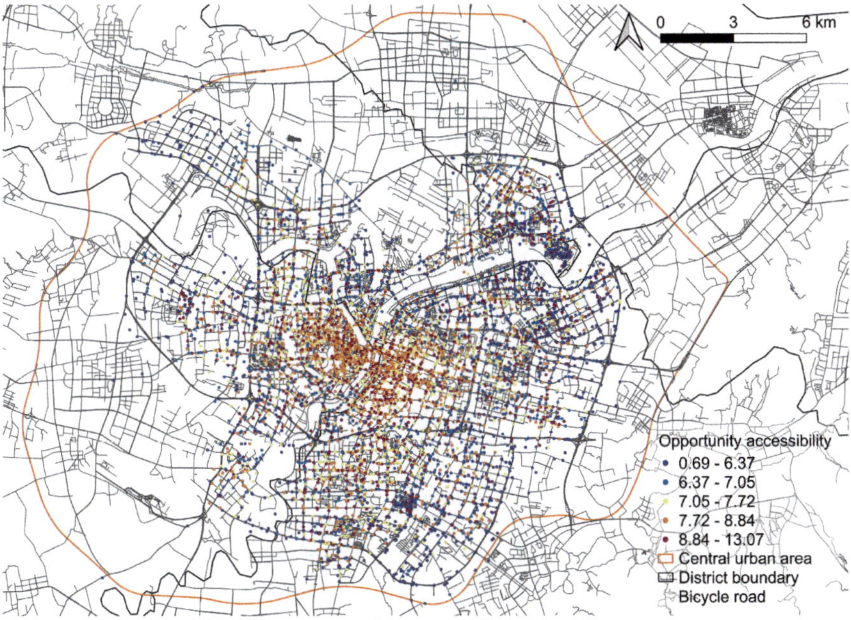

(b) The opportunity accessibility of geofences

Fig. 5.4 Spatial distributions and attributes of geofences presents the spatial distribution of the population size and the proportions of youth, elderly, females, and non-locals in each BAZ. The figure reveals that BAZs closer to the centre of the study area tend to have higher total populations, larger proportions of female and elderly residents, and a lower proportion of non-local residents. In contrast, the spatial distribution of the proportion of youth is more dispersed, with most BAZs showing a youth proportion of less than 1%.

5.5.3 Calculated Accessibility of BAZs

Based on the availability and opportunity accessibility of geofences and the population of BAZs, the SOA of BAZs is calculated using the KDBFCA method. The calculated SOA is then logarithmically transformed to concentrate the data, making it easier to analyse. The spatial distributions of SOA for BAZs under three distance thresholds are shown in Fig. 5.6. Similar to the opportunity accessibility of geofences, the SOA of BAZs follows a gradual decrease from the centre to the periphery of the study area. This pattern is understandable: compared to peripheral areas, the centre of the study area has more geofences with higher availability, greater access to opportunities, and a well-connected street network, as shown in Fig. 5.1b and Fig. 5.4. This infrastructure allows residents to reach these geofences within a reasonable walking distance.

By comparing the SOA of BAZs under three distance thresholds, we found that as the distance threshold increases, the average SOA of BAZs gradually improves, with values of 0.47, 0.49, and 0.50, respectively. Additionally, the differences in SOA among BAZs decrease, with values of 11.48, 11.11, and 10.42, respectively. This can be attributed to the fact that a larger distance threshold enables residents to access more geofences, thereby increasing the number of accessible POIs. Furthermore, expanding the catchment area raises the service demand at geofences, which, with a constant service capacity, reduces the service supply–demand ratio and the differences in the amount of service allocated to the BAZs.

5.5.4 Horizontal and Vertical Equity

This chapter evaluates horizontal equity by examining the distribution of DBS accessibility across BAZs using the Lorenz curve and Gini index. Figure 5.7 presents the Lorenz curves and Gini index results for the thresholds of 500 m, 750 m, and 1,000 m. The Gini index values for these three distance thresholds are 0.57, 0.52, and 0.49, respectively. These values suggest that the spatial distribution of accessibility still has room for improvement in terms of equity, as nearly 30% of the population lacks access to DBS. Furthermore, the Gini index demonstrates a negative correlation with the catchment area distance threshold, indicating that increasing the walking distance

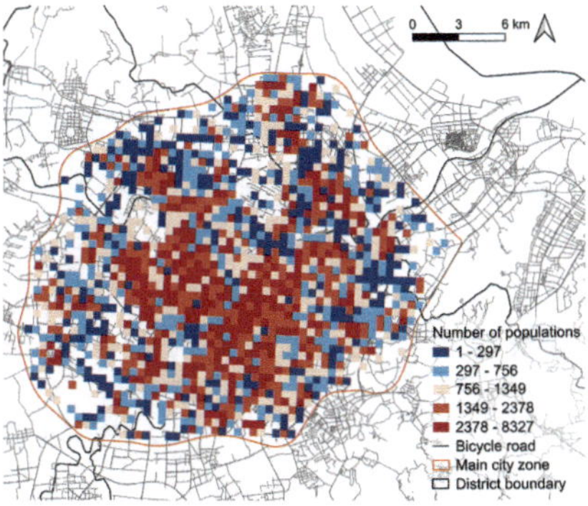

(a) Spatial distribution of the population number

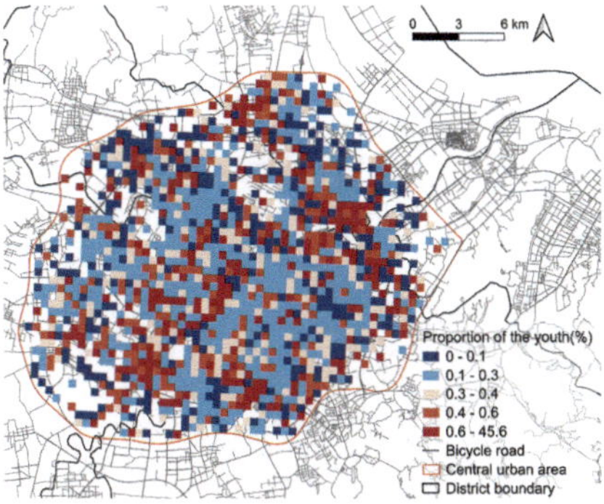

(b) Spatial distribution of the youth proportion

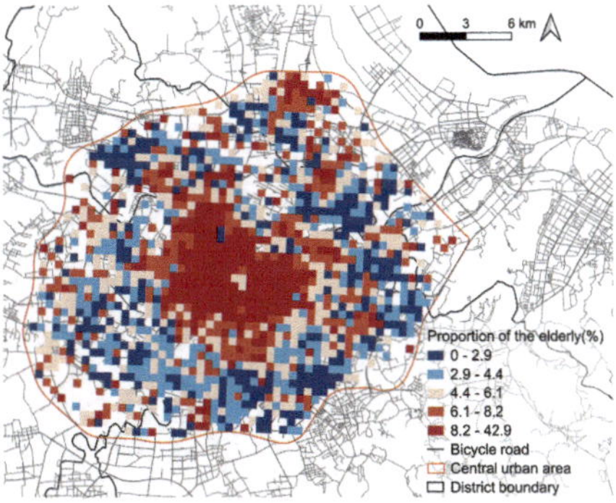

(c) Spatial distribution of the elderly proportion

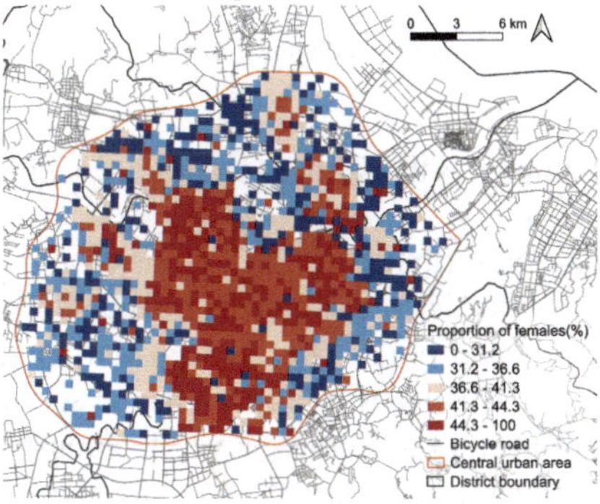

(d) Spatial distribution of the female proportion

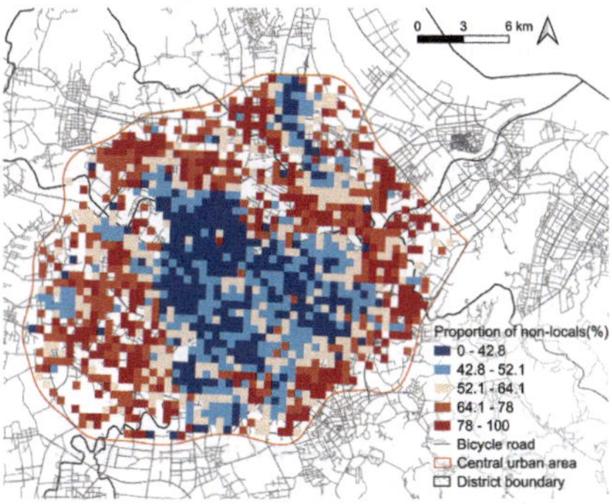

(e) Spatial distribution of the non-local proportion

Fig. 5.5 Spatial distributions of population and demographic attributes

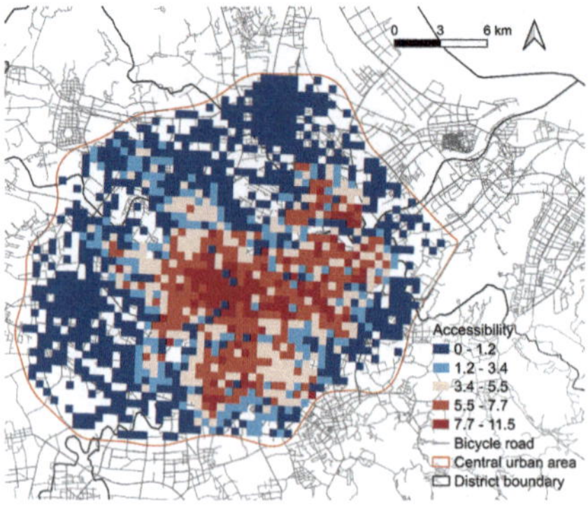

(a) The accessibility distribution (500m)

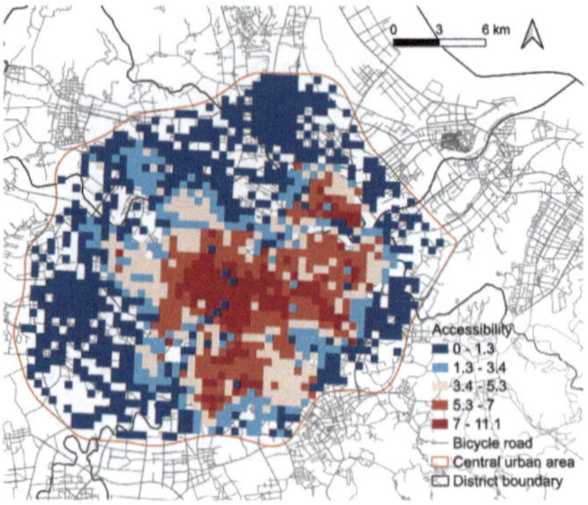

(b) The accessibility distribution (750m)

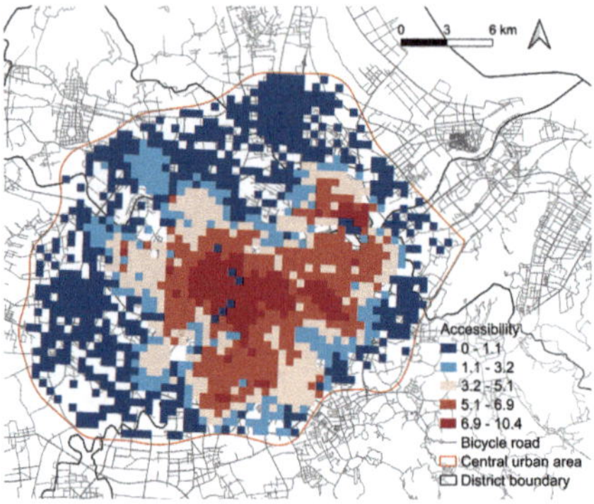

(c) The accessibility distribution (1000m)

Fig. 5.6 Spatial distributions of accessibility of BAZs under various distance thresholds

residents are willing to travel to access DBS improves the fairness of the accessibility distribution.

The decrease in the Gini index is primarily attributed to the expansion of the catchment area of facilities when the BFCA method is used to calculate accessibility. As the distance threshold increases, more demand points (i.e., BAZs) fall within the service range of facilities (i.e., geofences). This leads to an increase in service demand at facilities, while the supply–demand ratio decreases, given the constant service capacity of geofences. As a result, the variation in the amount of service allocated to each demand point is reduced. Consequently, the disparity in the SOA of BAZs is diminished, and the equity of SOA distribution improves.

The assessment of vertical equity is carried out using the KW test, Dunn's multiple comparison test, and Spearman's correlation coefficient to examine whether disadvantaged population groups have better accessibility. The study selects the proportion of the population under 18 (i.e., the youth), those over 65 (i.e., the elderly), females, and non-locals as indicators to measure the vulnerability level of BAZs. BAZs are

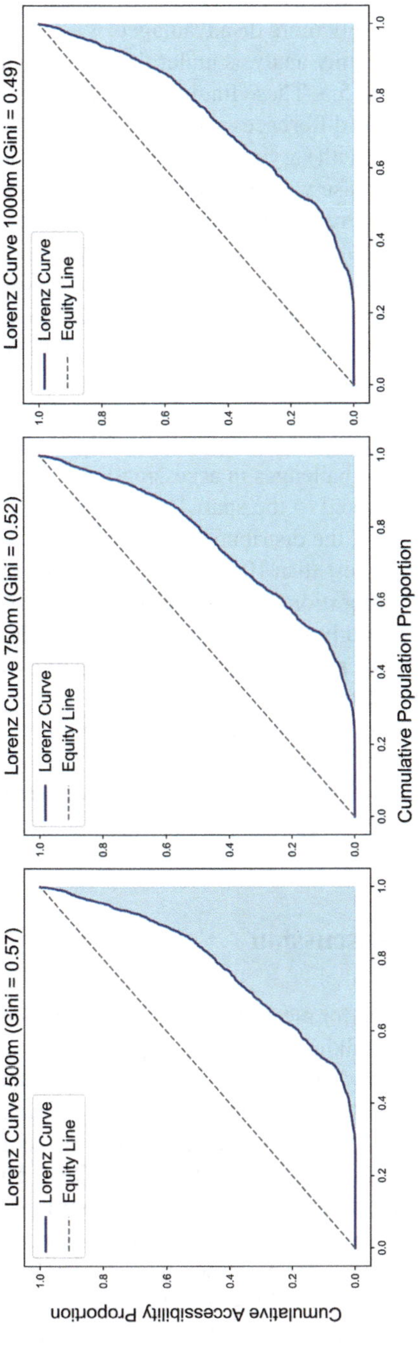

Fig. 5.7 Lorenz curves and Gini index

classified into five groups based on their SOA level using the Natural Breaks Classification method. A BAZ with a higher proportion of youth, elderly, females, and non-locals is considered relatively more disadvantaged.

The results of the vertical equity analysis under the three distance thresholds are presented in Tables 5.3, 5.4 and 5.5. These findings indicate that, with the exception of the youth, there are significant differences in the other three demographic attributes among the five groups of BAZs with varying SOA levels. Specifically, the proportions of the elderly and females increase with higher SOA levels, while the proportion of non-locals shows the opposite trend. The results of Dunn's test reveal significant differences in these three attributes between low-SOA BAZs and those with higher SOA levels. The Spearman correlation coefficients further support the significant relationships between these demographic attributes and SOA levels: the proportions of the elderly and females are positively correlated with SOA levels, whereas the proportion of non-locals is negatively correlated. This suggests that vertical equity issues exist in the distribution of SOA within the study area. BAZs with higher proportions of non-locals tend to have lower accessibility levels, highlighting that the non-local population faces challenges in accessing DBS.

These results are closely linked to the spatial distribution of various population attributes. As shown in Fig. 5.6, the distribution of the youth is relatively scattered, with over 90% of BAZs having less than 10% of their population in this age group. In contrast, BAZs with higher proportions of females and the elderly are more concentrated in the central area of the study area, which also exhibits higher SOA levels. This discrepancy may be due to lower mobile phone ownership rates among individuals under 18, leading to their underrepresentation in mobile phone data. Additionally, BAZs located in the peripheral areas tend to have higher proportions of non-locals. This may be a result of the non-local population generally having lower educational and income levels, and therefore residing in more affordable areas further from the city centre.

5.6 Conclusions and Discussion

This chapter proposes a method for measuring accessibility by integrating the availability of shared bikes with walking and riding opportunities to assess the equity performance of DBS in Ningbo, China, from both horizontal and vertical perspectives. Our findings indicate that central areas, with higher concentrations of population and economic activities, exhibit a greater density of geofences and higher opportunity accessibility, resulting in improved SOA for DBS. The DBS in Ningbo demonstrates better service levels in areas with higher proportions of females and the elderly, but lower levels in areas with a higher proportion of non-local populations. No significant correlation has been found between SOA and the youth. These findings contrast with studies in Western cities, which often report lower bike-sharing service levels in areas with higher densities of youth, elderly, and females [27, 43].

Table 5.3 The vertical equity results (500 m)

Group	SOA level	Median			
		Proportion of the youth (%)	Proportion of the elderly (%)	Proportion of females (%)	Proportion of non-locals (%)
1	Low	0.33	$4.64^{3***,4***,5***}$	$34.44^{2***,3***,4***,5***}$	$70.31^{2***,3***,4***,5***}$
2	Low–Medium	0.31	$4.88^{3**,4***,5***}$	$41.22^{4***,5***}$	$53.75^{4***,5*}$
3	Medium	0.33	5.81^{5***}	42.45^{4**}	49.78^{4*}
4	Medium–High	0.33	6.62^{5**}	43.72	45.42
5	High	0.35	9.38	43.85	46.61
KW test		3.90	104.64^{***}	412.66^{***}	272.23^{***}
Spearman's correlation coefficient		0.17	0.25^{***}	0.59^{***}	-0.47^{***}

Note: The medians of various demographic attributes for each group of BAZs are shown in columns 3–6 of rows 3–7 in the table. The superscript annotations indicate the results of Dunn's multiple comparison test. The numbers represent the group indices, and * denotes significance. For instance, the proportion of the elderly of BAZs in the low group is significantly different from that in the medium group. * indicates significance at the 0.05 level, ** indicates significance at the 0.01 level, and *** indicates significance at the 0.001 level

Table 5.4 The vertical equity results (750 m)

Group	SOA level	Median			
		Proportion of the youth (%)	Proportion of the elderly (%)	Proportion of females (%)	Proportion of non-locals (%)
1	Low	0.34	$4.55^{3***,4***,5***}$	$33.58^{2***,3***,4***,5***}$	$72.28^{2***,3***,4***,5***}$
2	Low–Medium	0.33	$4.93^{3**,4***,5***}$	$40.86^{3*,\ 4***,5***}$	$53.92^{4***,5***}$
3	Medium	0.30	5.81^{5***}	$42.45^{4*,5**}$	48.82
4	Medium–High	0.35	6.36^{5***}	43.50	46.26
5	High	0.35	9.47	44.04	45.13
KW test		3.90	4.58	125.19^{***}	487.53^{***}
Spearman's correlation coefficient		0.17	0.03	0.27^{***}	0.62^{***}

Table 5.5 The vertical equity results (1,000 m)

Group	SOA level	Median			
		Proportion of the youth (%)	Proportion of the elderly (%)	Proportion of females (%)	Proportion of non-locals (%)
1	Low	0.34	$4.57^{3^{**},4^{***},5^{***}}$	$33.24^{2^{***},3^{***},4^{***},5^{***}}$	$73.19^{2^{***},3^{***},4^{***},5^{***}}$
2	Low–Medium	0.31	$4.83^{4^{***},5^{***}}$	$39.59^{3^{***},4^{***},5^{***}}$	$56.00^{3^{**},4^{***},5^{***}}$
3	Medium	0.31	$5.25^{4^{***},5^{***}}$	$42.28^{5^{***}}$	$50.11^{4^{**},5^{**}}$
4	Medium–High	0.36	$6.56^{5^{***}}$	43.50	46.25
5	High	0.33	10.28	44.16	44.15
KW test		3.90	4.42	138.58^{***}	508.27^{***}
Spearman's correlation coefficient		0.17	0.03	0.27^{***}	0.62^{***}

Our results indicate that the distribution of DBS in Ningbo is inequitable, with residents in the peripheral areas of the study area and non-local populations facing disadvantages in accessing shared bikes. To address this issue, the following recommendations are proposed:

(1) Increasing the number of geofences and the availability of shared bikes in peripheral areas will enhance the accessibility of DBS services for disadvantaged populations, particularly non-local residents, living in these areas.
(2) Increasing investment in transport infrastructure to improve the walking environment and enhance the connectivity of cycling networks in areas with low accessibility will enable residents to easily reach more shared bikes and POIs within a reasonable distance, thereby improving the SOA of DBS.
(3) Public authorities can collaborate with bike-sharing companies to develop strategies aimed at improving the equity of their services. This could involve relaxing restrictions on the number of bikes deployed in underserved areas and reducing the usage costs of DBS services. Such collaborations would help ensure a more equitable distribution of DBS services across diverse populations.

The main contribution of this study lies in the development of an improved model for measuring DBS accessibility, which comprehensively incorporates the availability of shared bikes and urban opportunities. Using this model, we evaluate the equity performance of DBS in a Chinese urban context, supported by massive user-generated trip data, POIs, and mobile phone data. Our study provides valuable insights for urban policy-making and transport infrastructure investments related to DBS services.

A key limitation of this study is the use of mobile phone data to identify residents and retrieve their demographic attributes. While this approach demonstrates the potential of multi-source urban data to support the examination of the equity performance of DBS, especially in the absence of social information on DBS users due to privacy concerns, it may not fully capture the real demand for DBS services. This is because some identified residents may own private cars, bicycles, or electric bikes and may not rely on DBS services. Additionally, their demographic attributes (e.g., gender, age, and migration status) are limited, thus offering only partial profiles of disadvantaged population groups from a social perspective. Furthermore, the lack of shared bike users' attributes limits the calculation of trip preferences for different urban opportunities across various population groups. For example, a user with a job may have a stronger preference for using DBS to connect to public transit or workplaces.

Future research could aim to directly obtain socioeconomic information of DBS users through a questionnaire survey to enrich the accessibility evaluation and equity analysis. Comparing the equity performance of dockless bike-sharing and e-bike-sharing systems, which have gained popularity in recent years, would also be a valuable direction. Additionally, considering temporal variations in DBS trip preferences—such as the predominance of commuting trips on weekdays and recreational trips on weekends—could provide insights into the temporal differences in the accessibility and equity of urban DBS services.

References

1. Allen J, Farber S (2020) Planning transport for social inclusion: an accessibility-activity participation approach. Transp Res D Transp Environ 78:102212. https://doi.org/10.1016/j.trd.2019.102212
2. Campello RJGB, Moulavi D, Sander J (2013) Density-based clustering based on hierarchical density estimates. In: Pei J, Tseng VS, Cao L, Motoda H, Xu G (eds) Advances in knowledge discovery and data mining. Springer, Berlin, pp 160–172. https://doi.org/10.1007/978-3-642-37456-2_14
3. Camporeale R, Caggiani L, Ottomanelli M (2019) Modeling horizontal and vertical equity in the public transport design problem: a case study. Transp Res A Policy Pract 125:184–206. https://doi.org/10.1016/j.tra.2018.04.006
4. Chan Y, Walmsley RP (1997) Learning and understanding the Kruskal–Wallis one-way analysis-of-variance-by-ranks test for differences among three or more independent groups. Phys Ther 77:1755–1761. https://doi.org/10.1093/ptj/77.12.1755
5. Chen W, Chen X, Cheng L, Chen J, Tao S (2024) Locating new docked bike sharing stations considering demand suitability and spatial accessibility. Travel Behav Soc 34:100675. https://doi.org/10.1016/j.tbs.2023.100675
6. Chen Q, Ma S, Li H, Zhu N, He Q-C (2024) Optimizing bike rebalancing strategies in free-floating bike-sharing systems: an enhanced distributionally robust approach. Transp Res E Logist Transp Rev 184:103477. https://doi.org/10.1016/j.tre.2024.103477
7. Chen Y, Zhou L, Tang Y, Singh JP, Bouguila N, Wang C, Wang H, Du J (2019) Fast neighbor search by using revised *k*-d tree. Inf Sci 472:145–162. https://doi.org/10.1016/j.ins.2018.09.012
8. Di Ciommo F, Shiftan Y (2017) Transport equity analysis. Transp Rev 37:139–151. https://doi.org/10.1080/01441647.2017.1278647
9. Clockston RLM, Rojas-Rueda D (2021) Health impacts of bike-sharing systems in the U.S. Environ Res 202:111709. https://doi.org/10.1016/j.envres.2021.111709
10. Dai Y, Xu L, Zhang X, Fu Y, Dong W (2024) Promoting sustainable development: a study of China's bicycle sharing industry policies based on text analysis. Res Transp Bus Manag 52:101085. https://doi.org/10.1016/j.rtbm.2023.101085
11. Desjardins E, Higgins CD, Páez A (2022) Examining equity in accessibility to bike share: a balanced floating catchment area approach. Transp Res D Transp Environ 102:103091. https://doi.org/10.1016/j.trd.2021.103091
12. Duran-Rodas D, Villeneuve D, Pereira FC, Wulfhorst G (2020) How fair is the allocation of bike-sharing infrastructure? Framework for a qualitative and quantitative spatial fairness assessment. Transp Res A Policy Pract 140:299–319. https://doi.org/10.1016/j.tra.2020.08.007
13. Duran-Rodas D, Wright B, Pereira FC, Wulfhorst G (2021) Demand And/oR Equity (DARE) method for planning bike-sharing. Transp Res D Transp Environ 97:102914. https://doi.org/10.1016/j.trd.2021.102914
14. El-Geneidy A, Grimsrud M, Wasfi R, Tétreault P, Surprenant-Legault J (2014) New evidence on walking distances to transit stops: identifying redundancies and gaps using variable service areas. Transport 41:193–210. https://doi.org/10.1007/s11116-013-9508-z
15. Eren E, Uz VE (2020) A review on bike-sharing: the factors affecting bike-sharing demand. Sustain Cities Soc 54:101882. https://doi.org/10.1016/j.scs.2019.101882
16. Fu C, Zhu N, Ma S, Liu R (2022) A two-stage robust approach to integrated station location and rebalancing vehicle service design in bike-sharing systems. Eur J Oper Res 298:915–938. https://doi.org/10.1016/j.ejor.2021.06.014
17. Ge Y, Qu W, Qi H, Cui X, Sun X (2020) Why people like using bikesharing: factors influencing bikeshare use in a Chinese sample. Transp Res D Transp Environ 87:102520. https://doi.org/10.1016/j.trd.2020.102520
18. Geurs KT, van Wee B (2004) Accessibility evaluation of land-use and transport strategies: review and research directions. J Transp Geogr 12:127–140. https://doi.org/10.1016/j.jtrangeo.2003.10.005

19. Giannotti M, Tomasiello DB, Bittencourt TA (2022) The bias in estimating accessibility inequalities using gravity-based metrics. J Transp Geogr 101:103337. https://doi.org/10.1016/j.jtrangeo.2022.103337
20. Giuffrida N, Pilla F, Carroll P (2023) The social sustainability of cycling: assessing equity in the accessibility of bike-sharing services. J Transp Geogr 106:103490. https://doi.org/10.1016/j.jtrangeo.2022.103490
21. Guo S, Song C, Pei T, Liu Y, Ma T, Du Y, Chen J, Fan Z, Tang X, Peng Y, Wang Y (2019) Accessibility to urban parks for elderly residents: perspectives from mobile phone data. Landsc Urban Plan 191:103642. https://doi.org/10.1016/j.landurbplan.2019.103642
22. Hansen WG (1959) How accessibility shapes land use. J Am Inst Plan 25:73–76. https://doi.org/10.1080/01944365908978307
23. Hosford K, Fuller D, Lear SA, Teschke K, Gauvin L, Brauer M, Winters M (2018) Evaluation of the impact of a public bicycle share program on population bicycling in Vancouver, BC. Prev Med Rep 12:176–181. https://doi.org/10.1016/j.pmedr.2018.09.014
24. Huang Y, Liu W, Zhuo Y, Sun X (2020) Social insurance participation and urban settlement intentions of China's floating population. Popul Space Place 26:e2366. https://doi.org/10.1002/psp.2366
25. Hull Grasso S, Barnes P, Chavis C (2020) Bike share equity for underrepresented groups: analyzing barriers to system usage in Baltimore, Maryland. Sustainability 12:7600. https://doi.org/10.3390/su12187600
26. Ji S, Heinen E, Wang Y (2023) Non-linear effects of street patterns and land use on the bike-share usage. Transp Res D Transp Environ 116:103630. https://doi.org/10.1016/j.trd.2023.103630
27. Jin ST, Sui DZ (2024) Bikesharing and equity: a nationwide study of bikesharing accessibility in the U.S. Transp Res A Policy Pract 181:103983. https://doi.org/10.1016/j.tra.2024.103983
28. Kabra A, Belavina E, Girotra K (2020) Bike-share systems: accessibility and availability. Manag Sci 66:3803–3824. https://doi.org/10.1287/mnsc.2019.3407
29. Khashoggi BF, Murad A (2021) Use of 2SFCA method to identify and analyze spatial access disparities to healthcare in Jeddah, Saudi Arabia. Appl Sci 11:9537. https://doi.org/10.3390/app11209537
30. Kong H, Jin ST, Sui DZ (2020) Deciphering the relationship between bikesharing and public transit: modal substitution, integration, and complementation. Transp Res D Transp Environ 85:102392. https://doi.org/10.1016/j.trd.2020.102392
31. Li A, Gao K, Zhao P, Qu X, Axhausen KW (2021) High-resolution assessment of environmental benefits of dockless bike-sharing systems based on transaction data. J Clean Prod 296:126423. https://doi.org/10.1016/j.jclepro.2021.126423
32. Li A, Huang Y, Axhausen KW (2020) An approach to imputing destination activities for inclusion in measures of bicycle accessibility. J Transp Geogr 82:102566. https://doi.org/10.1016/j.jtrangeo.2019.102566
33. Li A, Zhao P, Huang Y, Gao K, Axhausen KW (2020) An empirical analysis of dockless bike-sharing utilization and its explanatory factors: case study from Shanghai, China. J Transp Geogr 88:102828. https://doi.org/10.1016/j.jtrangeo.2020.102828
34. Liang Z, Ma Z (2004) China's floating population: new evidence from the 2000 census. Popul Dev Rev 30:467–488. https://doi.org/10.1111/j.1728-4457.2004.00024.x
35. Lin JY-J, Jenelius E, Cebecauer M, Rubensson I, Chen C (2023) The equity of public transport crowding exposure. J Transp Geogr 110:103631. https://doi.org/10.1016/j.jtrangeo.2023.103631
36. Lin D, Zhang Y, Zhu R, Meng L (2019) The analysis of catchment areas of metro stations using trajectory data generated by dockless shared bikes. Sustain Cities Soc 49:101598. https://doi.org/10.1016/j.scs.2019.101598
37. Liu S, Zhu X (2004) Accessibility analyst: an integrated GIS tool for accessibility analysis in urban transport planning. Environ Plan B Plann Des 31:105–124. https://doi.org/10.1068/b305
38. Ma C, Zhang Y, Li Y, Wang Y, Jiang Y, Wang X, Ma S (2020) Healthcare, insurance, and medical expenditure of the floating population in Beijing, China. Front Public Health. https://doi.org/10.3389/fpubh.2020.00375

39. Millward H, Spinney J, Scott D (2013) Active-transport walking behavior: destinations, durations, distances. J Transp Geogr 28:101–110. https://doi.org/10.1016/j.jtrangeo.2012.11.012
40. Mix R, Hurtubia R, Raveau S (2022) Optimal location of bike-sharing stations: a built environment and accessibility approach. Transp Res A Policy Pract 160:126–142. https://doi.org/10.1016/j.tra.2022.03.022
41. Mohiuddin H, Fitch-Polse DT, Handy SL (2023) Does bike-share enhance transport equity? Evidence from the Sacramento, California region. J Transp Geogr 109:103588. https://doi.org/10.1016/j.jtrangeo.2023.103588
42. Mohiuddin H, Fitch-Polse DT, Handy SL (2024) Examining market segmentation to increase bike-share use and enhance equity: the case of the greater Sacramento region. Transp Policy 145:279–290. https://doi.org/10.1016/j.tranpol.2023.10.021
43. Mooney SJ, Hosford K, Howe B, Yan A, Winters M, Bassok A, Hirsch JA (2019) Freedom from the station: spatial equity in access to dockless bike share. J Transp Geogr 74:91–96. https://doi.org/10.1016/j.jtrangeo.2018.11.009
44. Padeiro M (2023) Cycling infrastructures and equity: an examination of bike lanes and bike sharing system in Lisbon, Portugal. Cities & Health
45. Paez A, Higgins CD, Vivona SF (2019) Demand and level of service inflation in floating catchment area (FCA) methods. PLoS ONE 14:e0218773. https://doi.org/10.1371/journal.pone.0218773
46. Pereira RHM, Schwanen T, Banister D (2017) Distributive justice and equity in transportation. Transp Rev 37:170–191. https://doi.org/10.1080/01441647.2016.1257660
47. Qian X, Jaller M (2021) Bikeshare destination choices and accessibility among disadvantaged communities. Transp Res D Transp Environ 91:102686. https://doi.org/10.1016/j.trd.2020.102686
48. Research, S.M., n.d. Stellar Market Research [WWW Document]. Stellar Market Research. https://www.stellarmr.com/report/North-America-Bike-Sharing-Market/50. Accessed 14 May 2024
49. Ross-Perez A, Walton N, Pinto N (2022) Identifying trip purpose from a dockless bike-sharing system in Manchester. J Transp Geogr 99:103293. https://doi.org/10.1016/j.jtrangeo.2022.103293
50. Saltykova K, Ma X, Yao L, Kong H (2022) Environmental impact assessment of bike-sharing considering the modal shift from public transit. Transp Res D Transp Environ 105:103238. https://doi.org/10.1016/j.trd.2022.103238
51. Song J, Zhang L, Qin Z, Ramli MA (2021) A spatiotemporal dynamic analyses approach for dockless bike-share system. Comput Environ Urban Syst 85:101566. https://doi.org/10.1016/j.compenvurbsys.2020.101566
52. Teixeira JF, Silva C, Moura e Sá F (2021) Empirical evidence on the impacts of bikesharing: a literature review. Transp Rev 41:329–351. https://doi.org/10.1080/01441647.2020.1841328
53. Tiznado-Aitken I, Fuenzalida-Izquierdo J, Sagaris L, Mora R (2021) Using the five Ws to explore bikeshare equity in Santiago, Chile. J Transp Geogr 97:103210. https://doi.org/10.1016/j.jtrangeo.2021.103210
54. Wang J, Lindsey G (2019) Do new bike share stations increase member use: a quasi-experimental study. Transp Res A Policy Pract 121:1–11. https://doi.org/10.1016/j.tra.2019.01.004
55. Wang C, Zhang Y (2023) Inferring and comparing trip purposes of shared micromobility services: a case study of Ningbo, China. Trans Urban Data Sci Technol 2:178–203. https://doi.org/10.1177/27541231231179884
56. Xing Y, Wang K, Lu JJ (2020) Exploring travel patterns and trip purposes of dockless bike-sharing by analyzing massive bike-sharing data in Shanghai. China. J Transp Geogr 87:102787. https://doi.org/10.1016/j.jtrangeo.2020.102787
57. Yen BTH, Mulley C, Zhang M (2020) Equity in financing public transport infrastructure: evaluating funding options. Transp Policy 95:68–77. https://doi.org/10.1016/j.tranpol.2020.06.001

58. Zhang Y, Lin D, Mi Z (2019) Electric fence planning for dockless bike-sharing services. J Clean Prod 206:383–393. https://doi.org/10.1016/j.jclepro.2018.09.215
59. Zhang Y, Mi Z (2018) Environmental benefits of bike sharing: a big data-based analysis. Appl Energy 220:296–301. https://doi.org/10.1016/j.apenergy.2018.03.101
60. Zhang X, Shen Y, Zhao J (2021) The mobility pattern of dockless bike sharing: a four-month study in Singapore. Transp Res D Transp Environ 98:102961. https://doi.org/10.1016/j.trd.2021.102961
61. Zhang M, Zhao P (2021) Literature review on urban transport equity in transitional China: from empirical studies to universal knowledge. J Transp Geogr 96:103177. https://doi.org/10.1016/j.jtrangeo.2021.103177

Chapter 6
Support the Development of 15-Minute Cities by Bike-Sharing Services

6.1 Introduction

Recently, the 15-min city model has received increasing attention in academic literature, policy-making and urban planning practices [23, 48]. It refers to a city where residents can conveniently access most of the essential amenities and services (e.g., healthcare, education, and groceries) within a given travel time from their homes [37]. It relies on seven basic principles, namely, human-scale urban design, density, diversity, flexibility, proximity, digitalisation, and connectivity [25]. Building 15-min cities is a way to support urban sustainable development. As a proximity-based planning approach, the 15-min city can help reduce car dependency, fuel consumption and environmental pollution [2]. Furthermore, walking within one's neighbourhood helps to increase social interactions and cohesion among residents [13, 30]. Accessing amenities and services via active transport is also associated with healthy benefits, such as lowering cardiovascular disease risk, reducing obesity and improving sleep quality [39, 49, 50]. It provides an opportunity to increase urban liveability and repair socio-economic fractures caused by contemporary urban planning methods [42].

At the 2015 Paris Climate Change Conference, the concept of the low-carbon city was introduced as a key policy agenda for large urban areas. One year later, Carlos Moreno, a Colombian-French urbanist, proposed the 15-min city model as a framework for tackling the climate change challenge [25, 37]. This model garnered significant attention primarily due to its potential to achieve a variety of urban policy objectives, including reducing car dependence, supporting local businesses, and promoting active travel habits [7, 31]. Its widespread appeal is demonstrated by the fact that Paris Mayor Anne Hidalgo advocated for transforming Paris into a 15-min city, while the C40 Cities Climate Leadership Group recommended it as a key strategy for supporting economic recovery from the pandemic [3, 36].

The variations of the model, however, were first proposed in Portland, USA, in 2008 when the city unveiled its vision for fostering 20-min neighbourhoods as part of its climate change mitigation strategy [34]. In 2016, Barcelona, Spain, introduced

Y. Zhang, *Understanding Shared Micromobility in China Using Massive User-Generated Trip Data*, Urban Sustainability,
https://doi.org/10.1007/978-981-96-8492-2_6

the superblock concept, aimed at reducing motorised traffic and creating space for people, active travel, and green areas [40]. Similarly, Melbourne, Australia, launched the Liveable Neighbourhoods programme as its approach to integrating the 20-min neighbourhood. Copenhagen, Denmark, proposed its "five-minutes-to-everything" principle. Sydney, Australia, has long prided itself on being a 30-min city, highlighting how it has enhanced the quality of life for its residents [29, 42]. The time threshold used to define the 15-min city (and its variations) ranges from 5 to 30 min, with 15 and 20 min being the most commonly adopted. Particularly, American and Australian cities, which tend to have lower population densities, favour the 20-min model, while 15 min is preferred by cities in Europe and Asia, where populations are typically denser than in Anglo-Saxon countries [48]. The 15-min timeframe in the model is not rigid and can be adapted based on specific urban contexts [16, 20, 37].

In China, the 15-min city is described using various terms, such as the "15-min living circle" and the "15-min life unit". For example, Shanghai released its planning guidance for 15-min community-life circles in August 2016. Additionally, 15-min living circles were incorporated and clarified in the National Standard for Planning and Designing Urban Residential Neighbourhoods, issued by the Ministry of Housing and Urban–Rural Development in 2018. Recently, more Chinese cities, including Beijing, Guangzhou, Hangzhou, and Wuhan, have introduced their own policies and planning strategies to support the development of 15-min cities.

Although the 15-min city (and its variations) has gained significant popularity in the past decade, researchers have argued that the model is not a radically new concept. It draws on long-established planning principles, such as the Garden City concept, the Neighborhood Unit, and the tenets of New Urbanism [44, 48]. By combining spatial and temporal dimensions, the model is grounded in the philosophy of chronourbanism, which posits that the quality of urban life is inversely proportional to the amount of time spent on transport [3, 37].

Bike-sharing, commonly provided by city governments or shared micromobility companies, is a shared service that offers short-term and affordable bike usage [5, 53]. Initially launched in the Netherlands in the 1960s, bike-sharing has been in existence for over half a century and has experienced significant growth in recent years. In contrast to traditional dock-based bike-sharing systems, which require users to borrow and return bikes at designated stations, dockless systems offer greater convenience, as there are no restrictions on parking bikes at fixed locations. As a sustainable mode of transport, bike-sharing is linked to various socioeconomic benefits, such as improved access to public transit, reduced greenhouse gas emissions, and the promotion of a healthy lifestyle [10, 12, 27, 53]. Due to these proven benefits, bike-sharing has gained increasing popularity, attracting substantial research attention in recent years [1, 43, 55].

When discussing the 15-min city model, researchers typically assume walking as the default mode of transport. However, studies that consider the 15-min city within a cycling context are limited [8, 26, 35]. With the support of cycling, including bike-sharing, residents' ability to reach essential amenities and services within 15 min can be significantly enhanced [32, 45]. Hosford et al. [20] found that Vancouver, the capital city of Canada, offers good cycling access to grocery stores, with less than

1% of the city's population unable to access a grocery store within a 15-min cycling distance. In contrast, around 20% of the population has no access to a local grocery store when walking. Another case in the United States also demonstrated cycling's potential to increase accessibility to various destinations for citizens [8].

Existing studies often assume that individuals can cycle directly from home to their destination. However, in the bike-sharing scenario, residents must first walk to the bike parking locations before riding to their desired amenities. As a result, the scope of a 15-min city with bike-sharing may be influenced by the distribution of shared bikes. Given the popularity of bike-sharing and its ability to enhance residents' access to local amenities and services, it is essential to evaluate how this emerging mode of mobility supports the development of 15-min cities.

This chapter is the first study to systematically examine the role of dockless bike-sharing services in the development of 15-min cities, comparing it with two other modes of transport (walking and cycling with private bikes). The study is conducted in Zhejiang, a province located in the southeastern part of China, and substantial comparisons are made between its cities of varying urban sizes. Additionally, this research fully leverages the advantages of multi-source urban data to quantify accessibility to amenities and services, primarily including shared bike trip big data, Points of Interest (POIs), and street networks sourced from OpenStreetMap.

6.2 Literature Review

Several issues related to the 15-min cities have been explored in existing studies, including the introduction of the conceptual model [37], assessment of its development status [33], examination of its underlying principles [19, 25], review of relevant planning policies [44], and optimisation of amenity locations [46]. In the following sections, we review the relevant literature from the perspectives of measurement, amenity selection, travel modes, and study areas.

Measuring 15-min cities is a fundamental issue in understanding the extent to which the model has been implemented in real cities. From a bottom-up perspective, Chen and Crooks [11] developed an agent-based model to simulate diverse communities and estimate their size to delineate 15-min cities using street network and POI data. Staricco [48] presented a method to measure which areas of the city of Turin (Italy) and what percentage of its population could access a given service within 5, 10, and 15 min by walking. Similarly, Logan et al. [34] critically examined the concept of the x-minute city and compared different approaches to measuring this concept by assessing accessibility in hundreds of urban areas in the United States and New Zealand. Birkenfeld et al. [6] identified which groups of households are living an x-minute city lifestyle to understand the compatibility of the planning approach in a North American context. They found that the 15- and 30-min city paradigms are goals that are difficult to achieve in Montréal (Canada), with only a few households enjoying proximity to meet their daily needs. From a socioeconomic perspective, certain population groups, such as ethnic minorities and lower-income residents, may face disadvantages in accessing key services and employment opportunities [21, 41]. In Paris (France), specific tools are required to identify the most disadvantaged neighbourhoods, ensuring an inclusive and enjoyable life experience within walking distance [3]. In this context, some scholars have focused on analysing the equity issues related to the 15-min city. Calafiore et al. [7] identified 20-min neighbourhoods in Liverpool (the United Kingdom) and assessed their association with social inequalities. In Utrecht (the Netherlands), Knap et al. [26] evaluated spatial and sociodemographic inequalities at the neighbourhood level. Gorrini et al. [17] conducted an extended spatial analysis to assess the walkability for children, who are particularly vulnerable to road accidents and require appropriate infrastructure to play outdoors and walk safely. Among these studies, some have adopted emerging data sources to support measurement, such as mobile phone data [54], POIs [11, 33], and street networks from OpenStreetMap [18]. However, many studies continue to rely on traditional surveys [6, 26, 34], which are typically collected in a labour-intensive and costly manner.

Regarding amenity selection, researchers have debated which amenities and services should be considered when measuring the 15-min city. Weng et al. [51] adopted categories such as education, commercial services, healthcare, municipal administration, finance and telecommunications, and elderly care to measure residents' preferences for amenities accessible within a walking distance of 5, 10, and 15 min in Shanghai, China. Moreno et al. [37] emphasised the need for six core urban functions: living, working, commerce, healthcare, education, and entertainment. Logan et al. [34] considered pharmacies, supermarkets, primary schools, and parks as key elements to inform relevant planning policies for creating a healthy city. Faith-based worship houses and services have also been recommended [8]. The categories of amenities considered vary across different urban contexts and are also constrained by factors such as data availability. Generally, amenity categories like education, health, and commerce (particularly food-related amenities) have been widely accepted [7, 24, 34].

In terms of travel modes, the 15-min city was initially proposed in research on walkability, which is considered a factor influencing urban life quality and accessibility [4]. Many studies have proposed or adopted walkability indicators to assess accessibility to amenities and services [14, 22]. As a result, walking is commonly regarded as the default travel mode for the 15-min city model [48]. In some cases, public transport is also considered when evaluating residents' accessibility [8, 26, 28]. Despite existing literature highlighting the importance of cycling in expanding residents' ability to meet essential needs with short trips [26], few studies have attempted to incorporate cycling into the investigation of the 15-min city [26]. For instance, McNeil [35] proposed a new bikeability indicator to assess how many services are accessible within the 20-min neighbourhoods in Portland (USA). Using Tempe (USA) as an example, Capasso Da Silva et al. [8] measured how destinations could be reached within 20 min by walking, cycling, and public transport. Hosford et al. [20] used a cumulative opportunity indicator to evaluate local accessibility to grocery stores by walking and cycling in Vancouver, Canada. Knap et al. [26] developed a new metric to assess how Utrecht (the Netherlands) adheres to the x-minute city model using cycling as the travel mode. To the best of our knowledge, the role of bike-sharing in developing the 15-min city has not been investigated or quantified in existing studies. This is the main research gap that we aim to address in this chapter.

In terms of study areas, most studies adopt a case-by-case approach, focusing on specific cities such as Montréal (Canada) [6], Naples (Italy) [15], New York City (the United States) [11], Paris (France) [37], Shanghai (China) [51], and Utrecht (the Netherlands) [26]. However, most of these cities are well-known large urban centres, and little attention has been given to small or medium-sized cities. Additionally, the majority of cities studied are located in Europe and North America, with cities from other continents, such as Asia, receiving less focus. Several studies have made comparisons between different cities. For instance, Logan et al. [34] conducted an evaluation based on the largest 500 cities in the United States at the census block level and all 43 urban areas in New Zealand at the Statistical Area level. Murgante et al. [38] assessed the implementation of the 15-min city in the Italian cities of Terni and Matera. However, multi-city comparisons are typically supported by traditional surveys and are predominantly conducted in Western countries. To date, no comparisons have been made across dozens of Chinese cities using emerging urban data sources, such as massive user-generated trip data.

6.3 Study Area and Data

6.3.1 Study Area

We have selected 26 district-level cities and 15 county-level cities in Zhejiang Province as the study area, as shown in Fig. 6.1. Zhejiang, located on the south-eastern coast of China, covers an area of 105,500 km^2 and has a population of 65.77 million as of 2022. It is one of the most economically developed provinces in China, with a gross domestic product of 8.26 trillion Yuan (approximately 1.15 trillion US dollars) in 2023. Zhejiang consists of 11 prefectures, including 37 district-level cities and 53 county-level cities. Unlike county-level cities, district-level cities tend to be closer to the central area of a prefecture-level city, with higher population densities and more advanced economic development. As both are at the same administrative level within China's urban system, it is appropriate to evaluate them jointly. The 41 selected cities have a total area of 40.69 km^2 and a combined population of 33.29 million. These cities are chosen for the study area because they are identified as having access to dockless bike-sharing services, based on the data obtained.

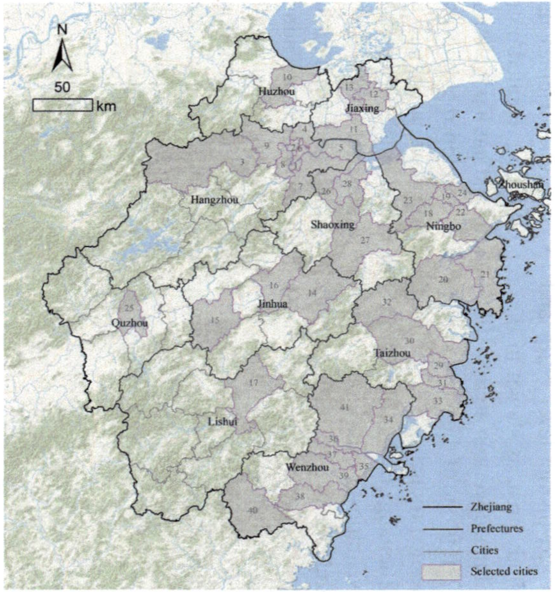

Prefectures	No	Selected cities	Prefectures	No	Selected cities
Hangzhou (HZ)	1	Binjiang	Ningbo (NB)	22	Yinzhou
	2	Gongshu		23	Yuyao
	3	Linan		24	Zhenhai
	4	Linping	Quzhou (QZ)	25	Kecheng
	5	Qiantang	Shaoxing (SX)	26	Keqiao
	6	Shangcheng		27	Shengzhou
	7	Xiaoshan		28	Yuecheng
	8	Xihu	Taizhou (TZ)	29	Jiaojiang
	9	Yuhang		30	Linhai
Huzhou (HuZ)	10	Wuxing		31	Luqiao
Jiaxing (JX)	11	Haining		32	Tiantai
	12	Nanhu		33	Wenling
	13	Xiuzhou	Wenzhou (WZ)	34	Yueqing
Jinhua (JH)	14	Dongyang		35	Longwan
	15	Wucheng		36	Lucheng
	16	Yiwu		37	Ouhai
Lishui (LS)	17	Liandu		38	Pingyang
Ningbo (NB)	18	Haishu		39	Ruian
	19	Jiangbei		40	Taishun
	20	Ninghai		41	Yongjia
	21	Xiangshan			

Fig. 6.1 Study area. Notes: Underlined names are county-level cities; others are district-level cities

6.3.2 *Data*

Major datasets used in this study include bike-sharing data, POIs, road networks and population data.

Bike-sharing data: This data is provided by Hellobike, one of the top three shared micromobility companies in China. After data preprocessing, the dataset covers 41 cities in Zhejiang Province, containing 12.31 million shared bike trips, generated by 2.13 million users and 381,910 bikes, between 14 and 27th September 2020. Each trip record includes the trip ID, user ID, start and end times, and the coordinates of the start and end locations. The average trip duration is 11.00 min, and the average trip distance is 1.52 km.

Points of Interest (POIs): The data is collected via the API (Application Programming Interface) of Gaode Maps, a leading map service provider in China. The original dataset contains 559,762 POIs across 43 mid-level categories, which are used to select essential amenities for the measurement of 15-min cities.

Road networks: The data is retrieved from OpenStreetMap using the OSMnx toolkit. Within the study area, the road network has a total length of 91,729.65 km, comprising 355,740 street segments and 244,890 street nodes.

Population data: The data is obtained from Worldpop, with a resolution of 1 km^2. It provides high-resolution data for mapping global human population distributions, which allows us to determine residential locations. In total, the study area has a population of 33.29 million residents.

Figure 6.2 shows the spatial distribution of the major datasets. As illustrated, bike usage, the number of POIs, and road density are positively correlated with population density. Population density, trip volume, POI numbers, and road density are higher in coastal areas (e.g., Longwan) and district-level cities (e.g., Xihu in Hangzhou).

6.4 The Analytical Framework

Figure 6.3 presents the analytical framework for 15-min cities, which consists of four steps: selecting amenities, identifying parking places, identifying 15-min units, and evaluating 15-min cities. Step 1 involves selecting amenities that are essential to the development of 15-min cities. In Step 2, this chapter identifies parking places for shared bikes by clustering the origins of trips. Step 3 further identifies 15-min units by calculating cumulative accessibility to essential amenities in three travel scenarios. Finally, in Step 4, the 15-min cities are evaluated from three perspectives: the coverage perspective, the amenity perspective, and the parking place perspective.

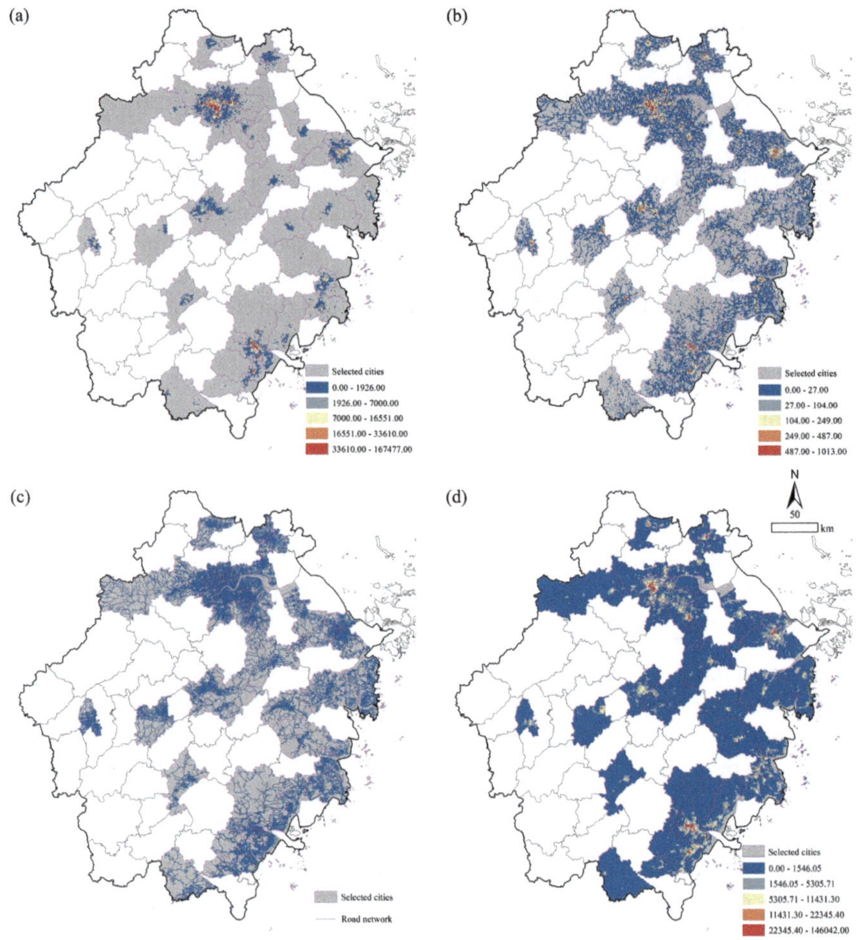

Fig. 6.2 Spatial distributions of the datasets. Note: (**a**) Bike-sharing trips (number/km^2); (**b**) POIs (number/km^2); (**c**) Road networks; (**d**) Population data (number/km^2)

6.4.1 Select Amenities

According to the National Standard of Planning and Designing Urban Residential Neighbourhoods (GB50180-2018) and prior literature [7, 51], this chapter defines ten essential categories of amenities: Public Transportation Sites, Schools, Cultural and Sports Facilities, Entertainment Venues, Financial Facilities, Hospitals, Pharmacies, Post Offices, Restaurants, and Shops. This chapter then merges the corresponding POIs with each category of amenities, as illustrated in Table 6.1.

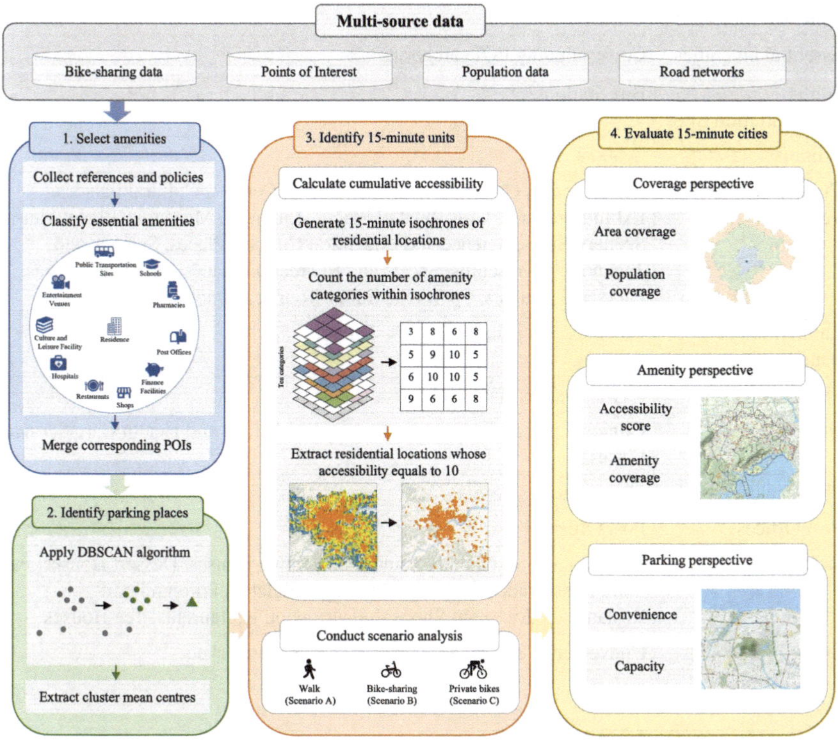

Fig. 6.3 The analytical framework

6.4.2 *Identify Parking Places*

Although dockless shared bikes can be parked in any authorised area, the company typically delivers or dispatches the bikes to specific parking places (e.g., locations near metro stations) with high travel demand. Therefore, it is reasonable to assume that users collect shared bikes from these parking places before their journeys. The Density-Based Spatial Clustering of Applications with Noise (DBSCAN) algorithm has demonstrated to be an effective method for identifying parking places for micromobility [47]. It requires only two parameters: the search radius of the neighbourhood (*Eps*) and the minimum number of trip origins (*MinPts*) within the *Eps*-neighbourhood. By setting these parameters, the DBSCAN algorithm can group points that are densely packed together (i.e., points with many nearby neighbours) into a cluster and identify points that lie alone in low-density regions as outliers (whose nearest neighbours are too far away). Each identified cluster can be considered a parking place, with the corresponding geographic mean centre being treated as the location of the parking place.

Table 6.1 Selected amenities

Essential amenities	Corresponding POI categories
Public Transportation Sites	Bus Stations
Schools	Schools
Culture and Sports Venues	Art Galleries, Convention and Exhibition Centres, Cultural Palaces, Exhibition Halls, Golf Related Places, Libraries, Museums, Planetariums, Scenery Spots, Science and Education Cultural Places, Science and Technology Museums, Sports and Recreation Places, Parks and Plazas, Parks and Squares, Sports Stadiums, Tourist Attraction Related
Entertainment Venues	Cinemas, Recreation Centres, Theatres
Finance Facilities	ATMs, Banks, Bank Related
Hospitals	Clinics, Emergency Centres, Hospitals, Medical and Health Care Service Places
Pharmacies	Pharmacies
Post Offices	Post Offices
Restaurants	Bakeries, Chinese Food Restaurants, Coffee Houses, Dessert Houses, Fast Food Restaurants, Food and Beverages Related, Foreign Food Restaurants, Icecream Shops, Leisure Food Restaurants, Tea Houses
Shops	Convenience Stores, Shopping Plazas, Supermarkets

6.4.3 Identify 15-Min Units

Different from existing literature, which often used real neighbourhoods or census tracts as the basic spatial units, this study regards each 1 km*1 km population grid as a residential location. A grid is defined as a 15-min unit (15MU) if its residents can access all categories of amenities within 15 min; otherwise, a non-15-min unit (non-15MU). To assess the diversity of the amenities that are accessible within 15 min, this chapter herein introduces a composite cumulative accessibility metric A_i, which represents the accessibility from the location of residential location i. Specifically, $A_i = \sum_{j=1}^{10} O_{ij}$, where O_{ij} is a binary value indicating whether amenity category j can be accessed by i. O_{ij} equals one if at least one amenity belonging to category j is accessible; otherwise, O_{ij} equals to 0. If A_i=10, i is identified as a 15MU.

In terms of travel mode, this chapter defines three distinctive scenarios that correspond to walking (Scenario A), bike-sharing (Scenario B), and cycling by private bikes (Scenario C). In Scenario B, residents have to walk to parking places and then ride a shared bike to amenities. Thus, the journey is a combination of walking and cycling. It is straightforward to capture the influence of bike-sharing on developing 15-min cities by comparing Scenario A and B. Scenario C can be regarded as an extreme form of bike-sharing in which everyone can access a shared bike at home, thus showing the potential of bike-sharing in improving the 15-min accessibility.

We proceed to generate 15-min isochrones from residential locations to calculate their cumulative accessibility in three scenarios. As shown in Fig. 6.4a, taking residents' locations as central points, the isochrones for Scenario A and C are depicted by the areas that are accessible within 15-min walking and cycling distances, respectively. The isochrone of Scenario B is additionally impacted by the locations of parking places. We first identify the parking places within 15-min walk of residents. Assuming the time to access a parking place is t_1 min, the remaining time for riding a shared bike is no more than $(15 - t_1)$ min. Therefore, the isochrone for each parking place is generated with a biking time of $(15 - t_1)$ min. By merging the isochrones of all parking places with the walking isochrone, the isochrone of Scenario B can be delineated.

Taking a residential location in Xihu (Hangzhou) as an example, Fig. 6.4b, c illustrate the processes of isochrone generation and accessibility calculation, respectively. The grid is adjacent to the West Lake Scenic Area and includes 4,549 residents. It shows the distribution of 194 POIs, with no Pharmacies or Post Offices within the 15-min walking isochrone. Therefore, the unit is classified as a non-15MU in Scenario A. Thirty-four parking places are identified as being accessible within a 15-min walking distance. Using these surrounding parking places as origins, their corresponding isochrones for riding shared bikes are delineated (shown in green). These new isochrones are merged with the walking isochrone (blue in the maps) to generate the isochrone for Scenario B. As a result, the unit can access 21 pharmacies and 11 post offices, thereby becoming a 15MU. The isochrone for Scenario C is created from the residents' location with a network-based distance of 3,000 m by private bike, and this range includes all categories.

6.4.4 *Evaluate 15-Min Cities*

The evaluation of 15-min cities is based on the 15-min units (15MUs), using three statistical perspectives: coverage, amenities, and parking places. The analysis spans three scenarios, with a particular focus on Scenario B. By comparing Scenario A with Scenario B, we can assess the additional benefits brought by bike-sharing. The comparisons between Scenario B and Scenario C are used to highlight the full potential of bike-sharing in shaping 15-min cities. Our evaluation aims to address three key questions. First, we assess the coverage of 15MUs in each city. What proportion of the area and population can access ten types of amenities within a 15-min journey? How does bike-sharing improve the coverage? Second, this chapter examines the diversity of amenities within 15-min cities, as well as the amenity coverage (the proportion of the area and population that can reach each type of amenity). How many types of amenities can be accessed within 15 min in each city? How does bike-sharing contribute to this and to the amenity coverage behaviour? Finally, we investigate the role of parking places in supporting the development of 15-min cities. Which parking places are crucial in transforming non-15MUs into

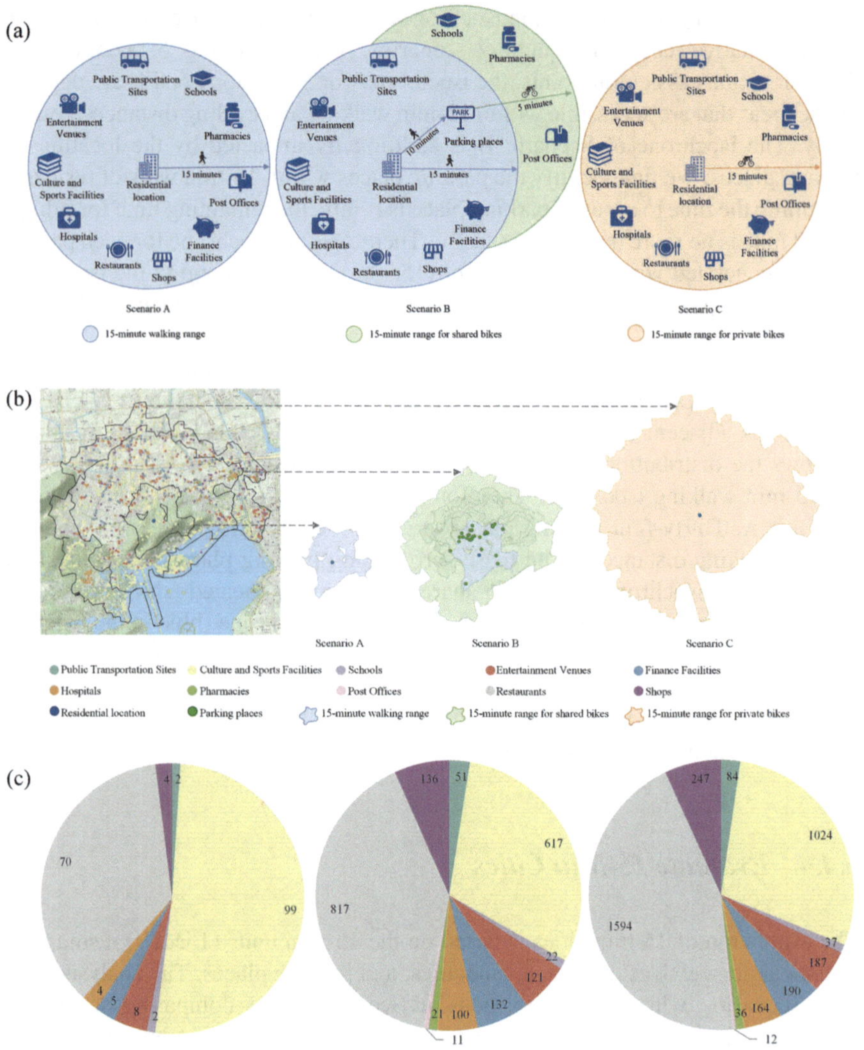

Fig. 6.4 Illustration of the 15-min unit identification. Note: (**a**) Illustration of three travel scenarios; (**b**) The grid's location, distribution of POIs and 15-min isochrones for three scenarios; (**c**) Number of amenity categories accessible in three scenarios

15MUs? How many residents benefit from these crucial parking places, and how convenient is access to them?

(1) Coverage perspective: The coverage-based analysis includes two indices: area coverage and population coverage. Area coverage measures the proportion of the total area covered by 15MUs at the city level. In contrast, population coverage represents the proportion of residents living within 15MUs. The formulas are as follows:

$$AC_k = \frac{S_{15MU_k}}{S_k} \tag{6.1}$$

$$PC_k = \frac{P_{15MU_k}}{P_k} \tag{6.2}$$

where, AC_k and PC_k represent the area coverage and population coverage of city k, respectively. S_{15MU_k} represents the area of 15-min units in city k; S_k represents the area of city k; P_{15MU_k} represents the population living in 15-min units in city k; and P_k represents the population of city k.

(2) Amenity perspective: The amenity-based analysis includes measures of accessibility scores and amenity coverage. Accessibility scores reflect the diversity of amenities, including population-weighted accessibility and the bottom 10% of accessibility. Population-weighted accessibility uses the population as a weight to measure accessibility across the 41 cities. To examine the accessibility of vulnerable groups, we further extract residential locations with the bottom 10% of accessibility in each city and measure their weighted accessibility.

$$PwACC_k = \sum_{i=1}^{N_k} \frac{P_{ik}}{P_k} A_{ik} \tag{6.3}$$

$$PwACC_k^{B10} = \sum_{i=1}^{N_k^{B10}} \frac{P_{ik}^{B10}}{P_k^{B10}} A_{ik}^{B10} \tag{6.4}$$

where, $PwACC_k$ and $PwACC_k^{B10}$ represent the weighted accessibility of the entire residential units and the residential units with the bottom 10% of accessibility in city k, respectively. P_{ik} is the population of the ith residential location in city k; A_{ik} is the accessibility score of the ith residential location in city k; N_k is the total number of residential locations in city k; P_{ik}^{B10} is the population of the ith residential location in the bottom 10% of accessibility in city k; P_k^{B10} represents the population in the residential locations in the bottom 10% of accessibility in city k; A_{ik}^{B10} is the accessibility score of the i-th residential location in the bottom of accessibility 10% in city k; and N_k^{B10} is the number of residential locations in the bottom 10% of accessibility in city k.

Amenity coverage illustrates the proportion of the area and population that can access at least one amenity within 15 min by category of amenities.

$$AAC_{jk} = \frac{S_{jk}}{S_k} \tag{6.5}$$

$$APC_{jk} = \frac{P_{jk}}{P_k} \tag{6.6}$$

where, AAC_{jk} and APC_{jk} represent the area coverage and population coverage to amenity j in city k, respectively. S_{jk} represents the area of units that can access amenity j in city k within 15 min; S_k represents the area of city k; P_j represents the population that can access amenity j within 15 min; and P_k represents the population of city k.

(3) Parking perspective: The parking-based analysis focuses on evaluating the convenience and capacity of crucial parking places. The parking places that help non-15MUs in Scenario A become 15MUs in Scenario B are selected. We then identify crucial parking places by calculating the travel time from these units (residential locations) to the corresponding parking places and then to the amenities. Crucial parking places are those that act as "transfer places" and minimise the total travel time. Figure 6.5 provides an example where two crucial parking places enable the residential location to access Entertainment Venues, Pharmacies, and Post Offices within 15 min.

The convenience of crucial parking places for a residential location is measured by the distance from the location to its corresponding crucial parking places. The shorter the distance, the more convenient the parking place is. The capacity of crucial parking places refers to the number of residents they serve. The more residents served, the larger the capacity of the crucial parking place. A parking place may serve as the crucial parking place for multiple units. Assuming a parking place l serve as the crucial parking place for N_{kl} residential units, then the convenience and capacity of this crucial parking place are measured as follows:

$$COcpk_{lk} = \frac{1}{n}\sum_{1}^{N_{kl}} d_{ij} \tag{6.7}$$

$$CAcpk_{lk} = \sum_{1}^{N_{kl}} P_{ik} \tag{6.8}$$

where, $COcpk_{lk}$ and $CAcpk_{lk}$ represent the convenience and capacity of crucial parking place l in city k, respectively. d_{ij} represents the distance between unit i and crucial parking place j, P_{ik} represents the population of unit i.

6.5 Results

6.5.1 Selected Amenities

This chapter selects a total of 559,762 amenities from POIs. For the sake of visualisation, the number of amenities in each category is transformed into a logarithmic scale and displayed using a radar chart (Fig. 6.6). As illustrated, the number of Restaurants (282.69 thousand) significantly exceeds those of other categories, accounting

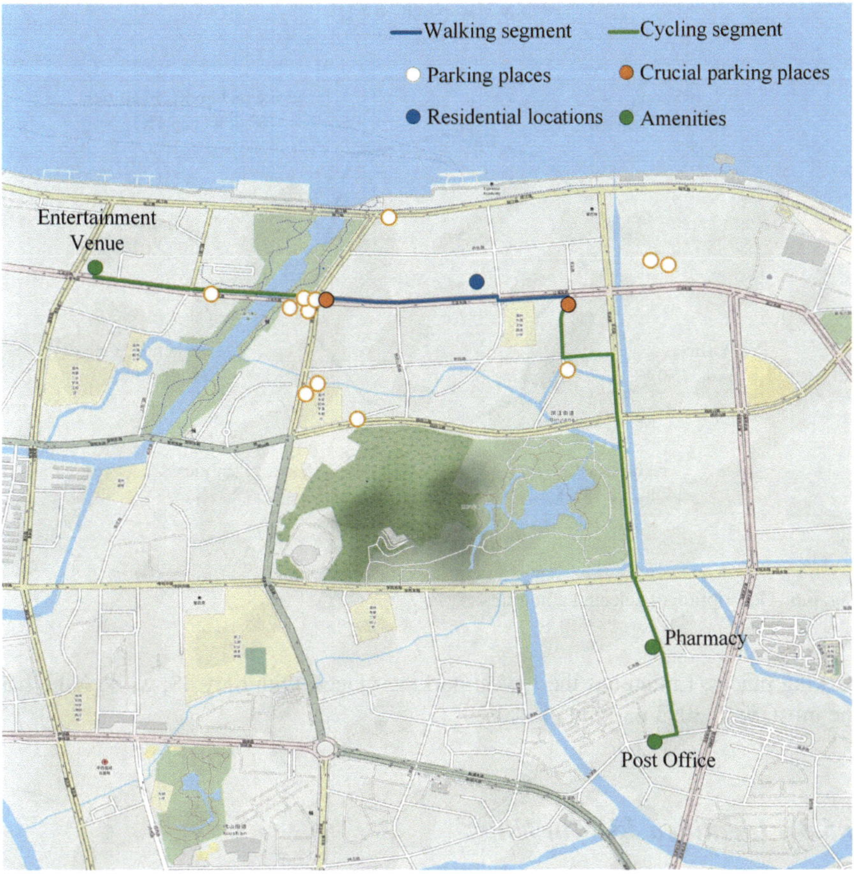

Fig. 6.5 Illustration of a crucial parking place

for 50.50% of the total amenities. There are also more Shops (69.10 thousand) and Culture and Sports Facilities (68.89 thousand) than other amenities. However, there are only 1.81 thousand Post Offices, indicating a low accessibility level for residents.

6.5.2 Identified Parking Places

We set *Eps* and *MinPts* in the DBSCAN algorithm to 3 m and 2 points, respectively. We identify 38,372 parking places, most of which are located in the central areas of the 41 cities (Fig. 6.7). Among the 45,822 grids in the selected cities, 3,292 grids have at least one parking place, with an average of 12 parking places per grid. Each parking place has an average of 23 daily trips, and each city has an average of 936

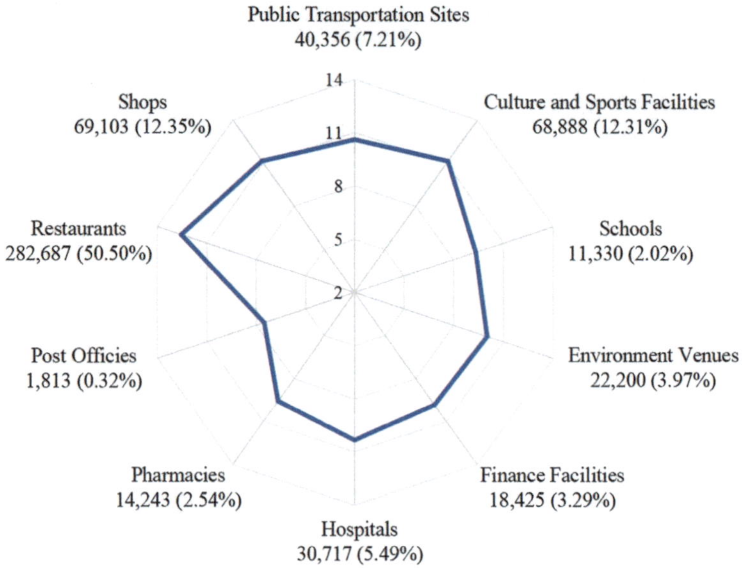

Fig. 6.6 Description of selected amenities

parking places. Linhai has the fewest parking places, with only 48, while Xihu has the most, with 3095 parking places.

6.5.3 Identified 15-Min Units

Using a walking speed of 80 m/min and a cycling speed of 200 m/min for shared bikes, the cumulative accessibility of the population grids in three scenarios is measured (Fig. 6.8a), and the corresponding 15-min units (15MUs) are identified subsequently (Fig. 6.8b). Among the 45,822 grids, 1,887 grids in Scenario A, 3,295 units in Scenario B, and 9,852 units in Scenario C are identified as 15MUs. The comparison between Scenario A and Scenario B highlights the accessibility improvements introduced by bike-sharing. The comparison between Scenario B and Scenario C underscores the significant role of parking places, as Scenario C can be considered a special form of bike-sharing where parking places are available at each residential location.

Taking a closer look at these spatial distributions, residents who can reach most types of essential amenities within 15 min predominantly reside in city centres or built-up areas, displaying a polycentric distribution pattern. Most units with low accessibility are located in the outer parts of cities, where fewer amenities and sparse road networks are found. The difference between city centres and urban fringes is more pronounced in Scenario B, which is attributed to the bike-sharing deployment scheme. Specifically, for commercial purposes, the service provider (i.e., Hellobike)

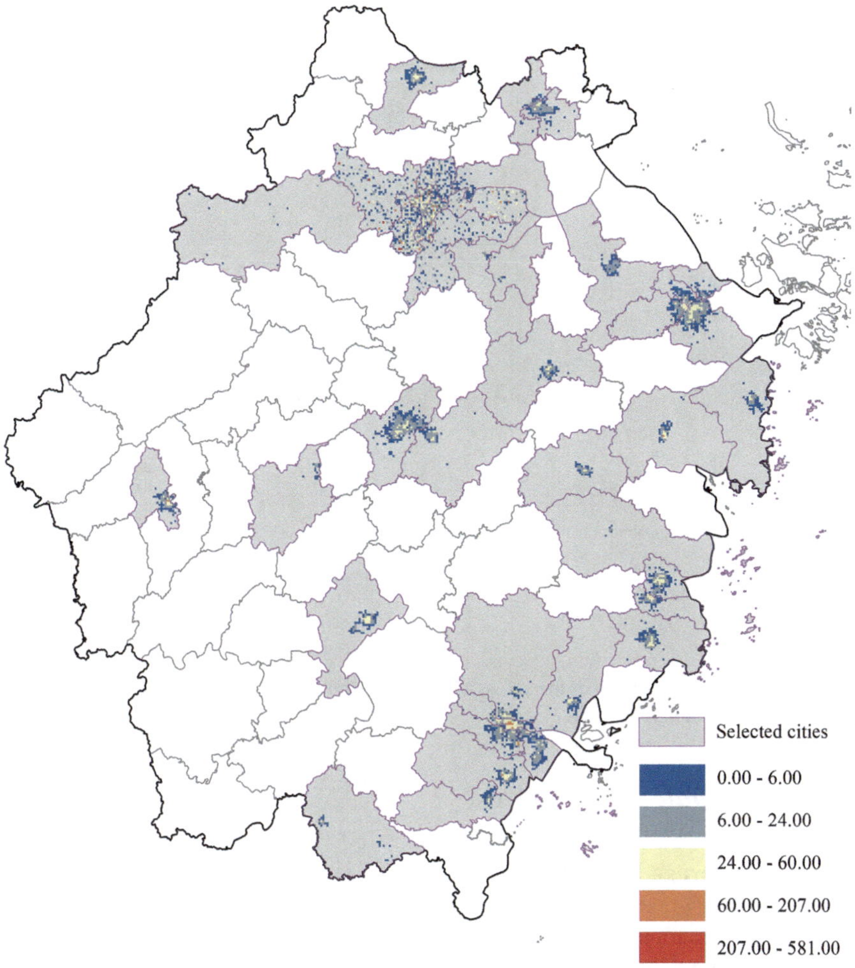

Fig. 6.7 The identified parking places. Note: The unit is number/km^2

deployed more shared bikes in the city centre, where accessibility was already higher. This unequal distribution leaves a large portion of the urban edges unserved by the bike-sharing system.

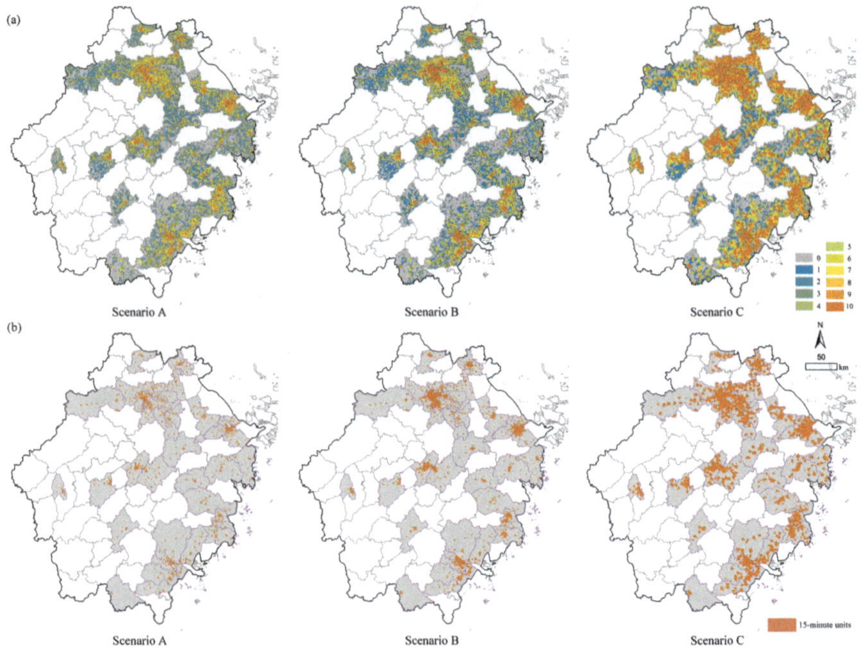

Fig. 6.8 Spatial distributions of the accessibility and 15-min units. Note: (**a**) Cumulative accessibility; (**b**) 15-min units

6.5.4 Evaluate 15-Min Cities

6.5.4.1 Coverage-Based Evaluation

In the three scenarios, the area coverage of 15MUs is 4.31% in Scenario A, 7.37% in Scenario B, and 22.34% in Scenario C, corresponding to 1.75, 3.00, and 9.09 thousand km^2, respectively. The proportions of residents living in 15MUs are 30.81%, 46.81%, and 73.79%, corresponding to 10.26 million, 15.58 million, and 24.56 million residents, respectively. The comparisons highlight the benefits and potential of bike-sharing in supporting 15-min cities. Moreover, the discrepancy between area coverage and population coverage suggests that most residents are concentrated in areas with smaller coverage. The coverage analysis also reveals different patterns at the city level (Fig. 6.9). In Scenario A, only three cities (Haishu, Gongshu, and Shangcheng) have over 50% of residents living in 15MUs. This proportion rises to 15 cities in Scenario B and 37 cities in Scenario C.

Regarding population coverage in Scenario B, the top five cities with the highest percentage of residents living in 15MUs are Gongshu (93.67%), Shangcheng (88.68%), Lucheng (76.72%), Binjiang (73.99%), and Jiangbei (71.62%). Conversely, the cities with the lowest percentages are Keqiao (15.73%), Linhai (17.93%), Pingyang (19.78%), Taishun (19.91%), and Linan (20.42%).

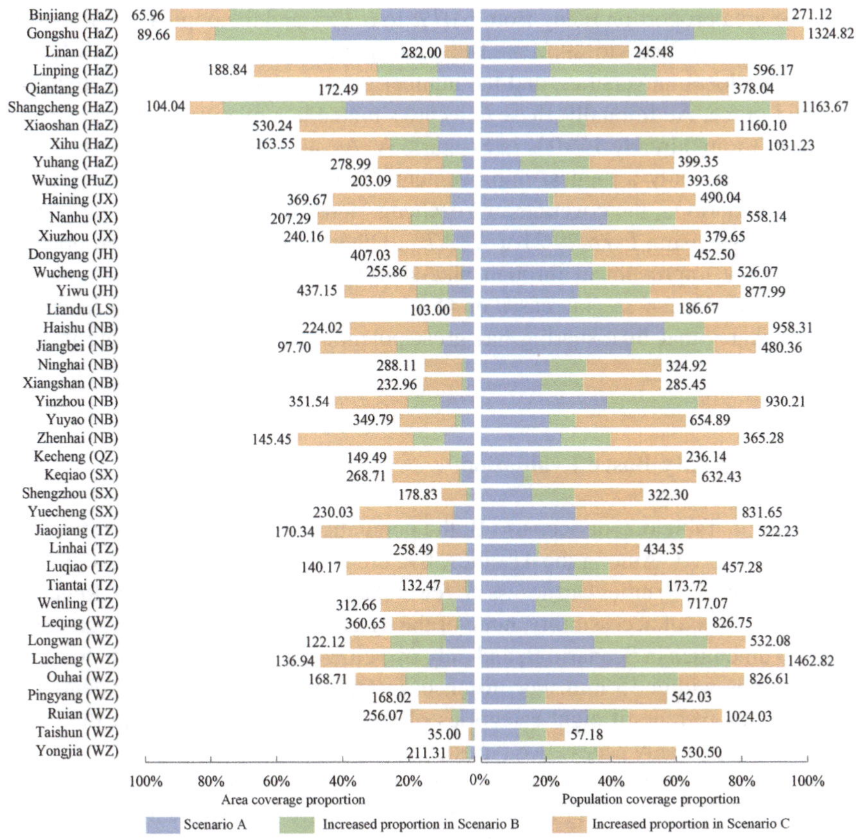

Fig. 6.9 Coverage of 15-min units in different cities. Note: The labels near the bars indicate the area of 15MUs (km^2) and the number of residents (thousand people) living in 15MUs in Scenario C, respectively

Generally, cities with better economic conditions tend to have higher coverage of 15MUs, while less developed cities have lower coverage. This result can be explained by two factors. First, developed areas attract more commercial activities and have more essential amenities. Second, the road networks in these areas are relatively well constructed, making it easier for residents to walk and cycle.

Among the 41 cities, Binjiang shows the highest coverage growth when bike-sharing is included in the calculation, with the area coverage expanding from 28.38 to 73.98% and the population coverage rising from 27.27 to 73.99%. Specifically, 134.93 thousand residents located in 32.54 km^2 have shifted from living in non-15MUs to 15MUs. The parking places in Binjiang are widely distributed throughout the entire city, and nearly all residents can access at least one parking place within 15 min by bike-sharing. As a result, bike-sharing is instrumental in expanding the 15-min cities in Binjiang. In contrast, Yuecheng shows the lowest increase in coverage, with the area proportion rising from 6.07% to 6.22% and the population proportion increasing from 29.07 to 29.21%. The parking places in Yuecheng are all located within two local universities, and the number of units that can access shared bikes is very limited. Additionally, these units are situated in a relatively prosperous area, where most are already 15MUs in Scenario A. As such, the introduction of shared bikes provides only a modest improvement. However, when comparing Scenario B with Scenario C, the coverage in Yuecheng increases significantly, with population coverage rising from 29.21 to 78.47%. Other cities with substantial potential for coverage improvement through bike-sharing include Xiaoshan, Haining, and Yueqing, which can see increases of 45.41%, 43.50%, and 40.61%, respectively. This indicates that bike-sharing has great potential to enhance coverage in these cities.

6.5.4.2 Amenity-Based Evaluation

As shown in Table 6.2, in Scenario B, the population-weighted accessibility ranges from 3.69 to 9.83. Among the 41 cities, five cities can access at least nine types of amenities, while 40 cities can access no fewer than five types of amenities. Regarding the bottom 10% of accessibility, none of the 41 cities can access an average of nine types of amenities in Scenario B, and only two cities can access more than five types of amenities within 15 min. For instance, in Ouhai and Longwan, residents have access to an average of eight categories of amenities, while those with the lowest accessibility can only access two types of amenities within 15 min.

By comparing the bottom 10% of accessibility in Scenario A with Scenario B, the introduction of bike-sharing improves the accessibility scores of vulnerable groups in 23 cities, primarily in those with high accessibility scores in Scenario A, such as Binjiang and Gongshu. However, in the remaining 18 cities, bike-sharing does not show significant improvements in the accessibility of vulnerable groups. When comparing Scenario B with Scenario C, we observe a marked improvement in most cities, highlighting the potential of bike-sharing to enhance accessibility for vulnerable groups. For example, Haishu and Linping show the greatest improvements. However, in Taishun, no improvement is observed due to the underdeveloped road network. This result suggests that the potential of bike-sharing to improve accessibility is limited by existing infrastructure. In other words, residents cannot expand their accessibility through cycling without a well-developed road network.

Figure 6.10a illustrates the coverage of various amenities under different scenarios. In Scenario B, 88.49% of residents can access at least one Public Transportation Site, 85.48% have access to Culture and Sports Venues, 84.50% can access Restaurants, and 84.43% can access Shops. However, Post Offices, a key daily service, still face accessibility challenges. In Scenario A, only 36.96% of residents can walk to a post office within 15 min. The population coverage of Post Offices in Scenario B increases to 51.09% and further rises to 77.59% in Scenario C. This suggests that shared bikes can effectively improve accessibility to Post Offices.

Table 6.2 Accessibility scores in different cities

Cities	Population-weighted accessibility			The bottom 10% of accessibility		
	Scenario A	Scenario B	Scenario C	Scenario A	Scenario B	Scenario C
Binjiang (HaZ)	8.11	9.25	9.86	2.66	4.24	8.56
Gongshu (HaZ)	9.44	9.83	9.98	6.49	7.99	9.79
Linan (HaZ)	5.25	5.30	7.81	0.18	0.18	1.32
Linping (HaZ)	7.76	8.49	9.75	2.58	3.02	8.29
Qiantang (HaZ)	6.84	7.92	9.23	0.51	0.70	4.06
Shangcheng (HaZ)	9.27	9.81	9.96	5.63	8.21	9.55
Xiaoshan (HaZ)	6.93	7.17	9.51	1.18	1.24	6.28
Xihu (HaZ)	8.41	9.00	9.74	2.59	3.26	7.73
Yuhang (HaZ)	6.37	6.93	9.14	0.57	0.73	5.00
Wuxing (HuZ)	6.53	6.83	8.88	0.44	0.44	3.16
Haining (JX)	6.29	6.37	9.30	0.90	0.95	6.06
Nanhu (JX)	7.33	7.78	9.39	0.71	0.74	5.24
Xiuzhou (JX)	5.94	6.08	8.94	0.01	0.01	4.00
Dongyang (JH)	6.15	6.28	8.59	0.05	0.06	1.71
Wucheng (JH)	6.94	7.07	9.12	0.42	0.42	3.45
Yiwu (JH)	6.88	7.50	9.28	0.56	0.72	4.21
Liandu (LS)	5.95	6.62	8.08	0.00	0.00	0.59
Haishu (NB)	8.29	8.52	9.68	1.56	1.64	7.01
Jiangbei (NB)	8.08	8.69	9.57	1.42	1.95	6.27
Ninghai (NB)	5.38	5.65	7.90	0.00	0.00	1.30
Xiangshan (NB)	5.38	5.72	8.17	0.00	0.00	1.60
Yinzhou (NB)	7.93	8.49	9.63	1.55	1.72	6.74
Yuyao (NB)	6.41	6.58	8.83	0.48	0.48	3.15
Zhenhai (NB)	6.61	6.94	9.25	0.06	0.14	4.05

(continued)

Table 6.2 (continued)

Cities	Population-weighted accessibility			The bottom 10% of accessibility		
	Scenario A	Scenario B	Scenario C	Scenario A	Scenario B	Scenario C
Kecheng (QZ)	5.27	5.96	8.72	0.00	0.00	2.99
Keqiao (SX)	6.45	6.51	9.15	1.00	1.00	4.65
Shengzhou (SX)	5.02	5.41	7.91	0.00	0.00	1.07
Yuecheng (SX)	7.26	7.31	9.29	1.05	1.05	4.46
Jiaojiang (TZ)	7.77	8.62	9.67	2.15	2.32	7.37
Linhai (TZ)	5.59	5.60	8.12	0.16	0.16	1.65
Luqiao (TZ)	7.01	7.36	9.13	0.72	0.73	3.91
Tiantai (TZ)	5.56	5.82	7.96	0.00	0.00	0.90
Wenling (TZ)	6.32	6.61	8.76	0.42	0.42	3.15
Leqing (WZ)	6.81	6.90	9.03	0.34	0.34	3.53
Longwan (WZ)	7.60	8.39	9.31	0.23	0.25	3.94
Lucheng (WZ)	8.56	9.31	9.80	2.89	4.38	7.98
Ouhai (WZ)	7.74	8.41	9.35	0.60	0.71	4.32
Pingyang (WZ)	6.09	6.32	8.84	0.33	0.34	3.04
Ruian (WZ)	7.33	7.64	9.14	0.51	0.55	3.67
Taishun (WZ)	3.43	3.69	5.90	0.00	0.00	0.00
Yongjia (WZ)	6.10	6.58	8.23	0.02	0.02	0.96

Overall, the results indicate that shared bikes can significantly enhance amenity accessibility, particularly for those amenities that are less accessible.

Figure 6.10b illustrates the coverage of various amenities that residents in the bottom 10% of accessibility can reach within 15 min. In Scenario B, Public Transportation Sites cover 35.02% of the bottom 10% of residents, followed by Culture and Sports Facilities (29.07%), Restaurants (20.86%), and Shops (18.48%). In Scenario A, Post Offices cover only 0.34% of the bottom 10% of residents, indicating a significant accessibility gap. In Scenario B, the coverage of Post Offices increases to 1.59%, and further rises to 11.40% in Scenario C. This suggests that shared bikes can significantly improve accessibility for these residents. It also indicates that shared bikes can not only enhance overall amenity accessibility but also considerably increase

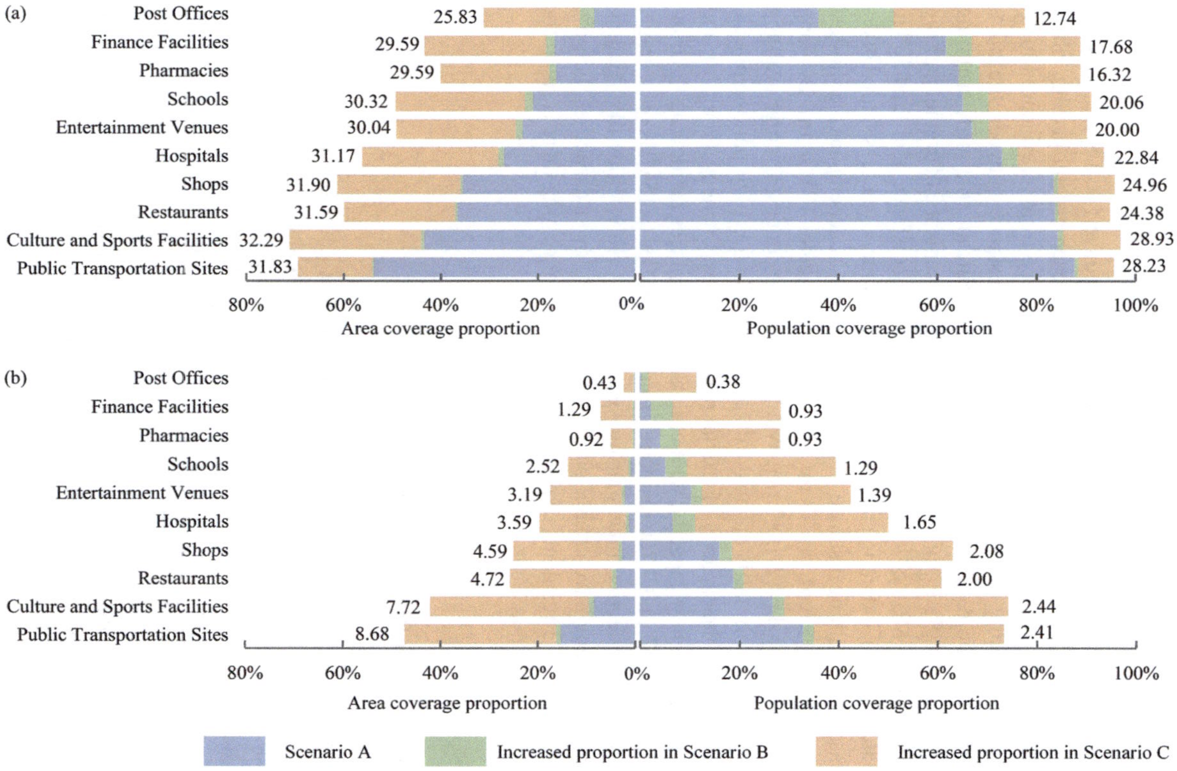

Fig. 6.10 Coverage of the ten amenities for all residents and residents with the bottom 10% of accessibility. Note: The labels near the bars indicate the area of units (km^2) and the number of residents (thousand people) who can access the amenities within 15 min in Scenario C

coverage for vulnerable groups, contributing to a more equitable and efficient 15-min city.

6.5.4.3 Parking-Based Evaluation

In total, we identify 1,458 crucial parking places that convert 1,408 non-15MUs in Scenario A into 15MUs in Scenario B. On average, each crucial parking place benefits 3,714 residents, with the average walking distance from a residential location to a crucial parking place being 286.54 m. The distance between the bike-sharing station and the residential location plays a significant role in determining bike-sharing adoption [9]. Figure 6.11a illustrates the convenience of crucial parking places in the 41 cities. Most cities have an average walking distance of around 300 m (approximately 4 min), indicating that residents can conveniently access shared bikes within a reasonable walking distance. Among the 41 cities, Yuecheng has the shortest average walking distance of 149.77 m, followed by Qiantang (173.38 m), Binjiang (192.33 m), and Gongshu (210.89 m). Wucheng has the longest average walking distance of 482.27 m, primarily due to the dispersed distribution of its parking places. Figure 6.11b presents the capacity of crucial parking places. Lucheng, Gongshu, and Wucheng have the highest capacity, with crucial parking places facilitating the transition of 10,726, 8,297, and 7,760 residents from non-15MUs to 15MUs on average, indicating a highly effective bike-sharing system. Conversely, cities like Tiantai, Taishun, and Dongyang have crucial parking places with lower capacities, helping 1,968, 1,925, and 1,911 residents transition on average, suggesting there is room for improvement in their bike-sharing infrastructure.

6.6 Conclusions and Discussion

First proposed in 2016, the concept of 15-min cities has gained significant popularity in recent years. Recognising the value of this concept, the Chinese government launched the Three-Year Action Plan to Comprehensively Promote the Construction of Urban Quarter-Hour Convenient Living Areas (2023–2025). This plan prioritises the allocation of shopping, catering, and other basic amenities to make daily life more convenient for residents, right on their doorsteps. As an emerging micromobility service, bike-sharing serves as an attractive tool for supporting such short-distance travel. Given that not every household owns a private bike, coupled with the unprecedented popularity of shared bikes in China, bike-sharing has become a compelling option for the creation of 15-min cities. In this chapter, we propose a methodological framework to systematically measure the 15-min city across three scenarios (walking, shared bikes, and private bikes) and apply it to Zhejiang Province. We identified 1,887 grids in Scenario A, 3,295 grids in Scenario B, and 9,852 grids in Scenario C as 15MUs. The corresponding 15MU populations are 10.26 million,

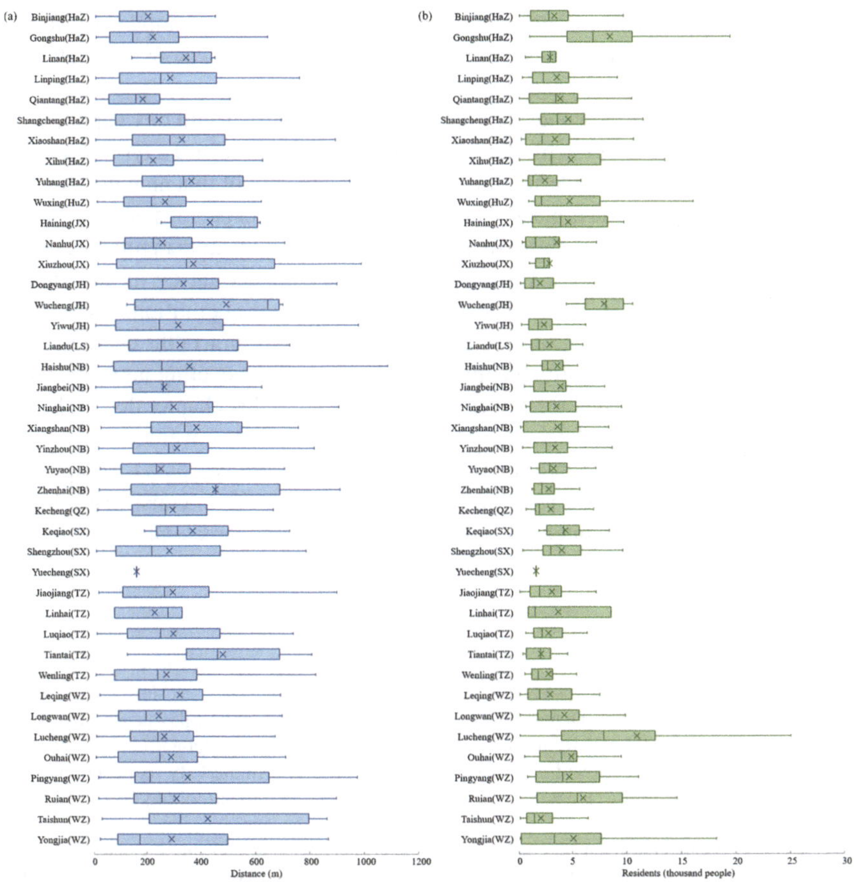

Fig. 6.11 The convenience (a) and capacity (b) of the crucial parking places in 41 cities

15.58 million, and 24.56 million, respectively. The results suggest that shared bikes are an effective means of supporting the development of 15-min cities.

Residents who can access most types of essential amenities within 15 min predominantly live in city centres or built-up areas, exhibiting a polycentric distribution pattern. The difference between city centres and urban fringes is more pronounced in Scenario B. This can be attributed to the fact that businesses deploy more shared bikes in city centres, where accessibility is higher. Furthermore, this unequal distribution leaves a large portion of the study area on the urban edge unserved by the bike-sharing system. The disparity becomes more uniform in Scenario C, highlighting the potential of bike-sharing to improve accessibility for vulnerable groups. Policymakers should focus more on residents in areas with low accessibility and limited improvement potential, thus avoiding structural inequalities. However, in practice, there are additional challenges in using shared bikes to promote 15-min cities. For instance, the objectives of bike-sharing companies and urban transport authorities are not always aligned. This misalignment may occur when private companies prioritise higher revenue and reduce the supply of shared bikes in areas where affordable and convenient connections are needed, yet are not profitable.

This chapter provides a starting point for identifying a 15-min city supported by shared bikes, although some limitations remain. Firstly, we regard a type of amenity as accessible without considering the number and capacity of that amenity. It would be valuable to explore the impact of amenity numbers and capacities on accessibility measurements and to adopt more advanced accessibility indicators in future research. Secondly, the definition of a 15-min city may vary depending on an individual's preferences and lifestyle. It would be interesting to adjust the definition of the 15-min city according to the cultural and social context of the study area or the characteristics of specific user groups. Lastly, the quality and safety of the cycling network, which may affect the actual travel time, were not considered in this chapter due to the lack of detailed environmental factors. For the sake of calculation convenience, we used a standard biking speed (12 km/h) to measure cycling distance. We plan to design more sophisticated models to assess biking speed in future research, considering factors such as motorised traffic intensity, cycling lanes, and slope gradients.

References

1. Alimo PK, Agyeman S, Danesh A, Yu C, Ma W (2023) Is public bike-sharing feasible in Ghana? Road users' perceptions and policy interventions. J Transp Geogr 106:103509
2. Allam Z, Bibri SE, Chabaud D, Moreno C (2022) The '15-Minute City' concept can shape a net-zero urban future. Human Soc Sci Commun 9(1):126. https://doi.org/10.1057/s41599-022-01145-0
3. Allam Z, Nieuwenhuijsen M, Chabaud D, Moreno C (2022) The 15-minute city offers a new framework for sustainability, liveability, and health. Lancet Planet Health 6(3):e181–e183. https://doi.org/10.1016/S2542-5196(22)00014-6
4. Barton H, Horswell M, Millar P (2012) Neighbourhood accessibility and active travel. Plan Pract Res 27(2):177–201
5. Bi H, Ye Z, Zhu H (2024) Mining bike sharing trip record data: a closer examination of the operating performance at station level. Transportation 51(3):1015–1041. https://doi.org/10.1007/s11116-022-10342-4
6. Birkenfeld C, Victoriano-Habit R, Alousi-Jones M, Soliz A, El-Geneidy A (2023) Who is living a local lifestyle? Towards a better understanding of the 15-minute-city and 30-minute-city concepts from a behavioural perspective in Montréal, Canada. J Urban Mobil 3:100048. https://doi.org/10.1016/j.urbmob.2023.100048
7. Calafiore A, Dunning R, Nurse A, Singleton A (2022) The 20-minute city: an equity analysis of Liverpool City Region. Transp Res Part D: Transp Environ 102:103111
8. Capasso Da Silva D, King DA, Lemar S (2020) Accessibility in practice: 20-minute city as a sustainability planning goal. Sustainability 12(1):129
9. Chen Z, Guo Y, Stuart AL, Zhang Y, Li X (2019) Exploring the equity performance of bike-sharing systems with disaggregated data: a story of southern Tampa. Transp Res Part A Policy Pract 130:529–545. https://doi.org/10.1016/j.tra.2019.09.048
10. Chen Y, Zhang Y, Coffman D, Mi Z (2022) An environmental benefit analysis of bike sharing in New York City. Cities 121:103475
11. Chen Q, Crooks AT (2021) Delineating a '15-minute city': an agent-based modeling approach to estimate the size of local communities. In: Proceedings of the 4th ACM SIGSPATIAL

international workshop on geospatial simulation, pp 29–37. https://doi.org/10.1145/3486184.3491080
12. Clockston RLM, Rojas-Rueda D (2021) Health impacts of bike-sharing systems in the US. Environ Res 202:111709
13. Dempsey N, Bramley G, Power S, Brown C (2011) The social dimension of sustainable development: defining urban social sustainability. Sustain Dev 19(5):289–300
14. D'Orso G, Migliore M (2020) A GIS-based method for evaluating the walkability of a pedestrian environment and prioritised investments. J Transp Geogr 82:102555
15. Gaglione F, Gargiulo C, Zucaro F, Cottrill C (2022) Urban accessibility in a 15-minute city: a measure in the city of Naples, Italy. Transp Res Procedia 60:378–385. https://doi.org/10.1016/j.trpro.2021.12.049
16. Gao W, Cui M, Pan E, Loo BP (2024) Green commuting within the x-minute city: towards a systematic evaluation of its feasibility. J Transp Geogr 121:104003
17. Gorrini A, Presicce D, Messa F, Choubassi R (2023) Walkability for children in Bologna: beyond the 15-minute city framework. J Urban Mobil 3:100052. https://doi.org/10.1016/j.urbmob.2023.100052
18. Graells-Garrido E, Serra-Burriel F, Rowe F, Cucchietti FM, Reyes P (2021) A city of cities: measuring how 15-minutes urban accessibility shapes human mobility in Barcelona. PLoS ONE 16(5):e0250080. https://doi.org/10.1371/journal.pone.0250080
19. Guzman LA, Oviedo D, Cantillo-Garcia VA (2024) Is proximity enough? A critical analysis of a 15-minute city considering individual perceptions. Cities 148:104882
20. Hosford K, Beairsto J, Winters M (2022) Is the 15-minute city within reach? Evaluating walking and cycling accessibility to grocery stores in Vancouver. Transp Res Interdiscip Perspect 14:100602. https://doi.org/10.1016/j.trip.2022.100602
21. Hu L (2019) Racial/ethnic differences in job accessibility effects: explaining employment and commutes in the Los Angeles region. Transp Res Part D: Transp Environ 76:56–71
22. Jardim B, de Castro Neto M, Barriguinha A (2023) A street-point method to measure the spatiotemporal relationship between walkability and pedestrian flow. Comput Environ Urban Syst 104:101993
23. Jiang J, Qiao W, Chuang I-T, Li Y, Wang T, Beattie L (2024) Mapping liveability: the "15-min city" concept for car-dependent districts in Auckland, New Zealand. Appl Geogr 163:103197
24. Kesarovski T, Hernández-Palacio F (2023) Time, the other dimension of urban form: measuring the relationship between urban density and accessibility to grocery shops in the 10-minute city. Environ Plan B: Urban Anal City Sci 50(1):44–59. https://doi.org/10.1177/23998083221103259
25. Khavarian-Garmsir AR, Sharifi A, Sadeghi A (2023) The 15-minute city: urban planning and design efforts toward creating sustainable neighborhoods. Cities 132:104101. https://doi.org/10.1016/j.cities.2022.104101
26. Knap E, Ulak MB, Geurs KT, Mulders A, Van Der Drift S (2023) A composite X-minute city cycling accessibility metric and its role in assessing spatial and socioeconomic inequalities – A case study in Utrecht, the Netherlands. J Urban Mobil 3:100043. https://doi.org/10.1016/j.urbmob.2022.100043
27. Kong H, Jin ST, Sui DZ (2020) Deciphering the relationship between bikesharing and public transit: modal substitution, integration, and complementation. Transp Res Part D: Transp Environ 85:102392
28. Larsson A, Elldér E, Vafeiadis E, Curtis C, Steiner A (2022) Exploring the potential for sustainable accessibility across settlement types. A Swedish case. Transp Res Part D Transp Environ 107:103297. https://doi.org/10.1016/j.trd.2022.103297
29. Leao, S., Hassan, M., Rashidi, T., & Pettit, C. (2021). Is Sydney a 30-minute city? Big data analytics assists to bring political rhetoric into practice. In M. Birkin, G. Clarke, J. Corcoran, & R. Stimson (Eds.), Big data applications in geography and planning: An essential companion. Edward Elgar Publishing.
30. Leyden KM (2003) Social capital and the built environment: the importance of walkable neighborhoods. Am J Public Health 93(9):1546–1551

31. Li Y, Chai Y, Chen Z, Li C (2023) From lockdown to precise prevention: adjusting epidemic-related spatial regulations from the perspectives of the 15-minute city and spatiotemporal planning. Sustain Cities Soc 92:104490. https://doi.org/10.1016/j.scs.2023.104490
32. Lin D, Zhang Y, Zhu R, Meng L (2019) The analysis of catchment areas of metro stations using trajectory data generated by dockless shared bikes. Sustain Cities Soc 49(8):101598
33. Liu D, Kwan M-P, Wang J (2024) Developing the 15-minute city: a comprehensive assessment of the status in Hong Kong. Travel Behav Soc 34:100666
34. Logan TM, Hobbs MH, Conrow LC, Reid NL, Young RA, Anderson MJ (2022) The x-minute city: measuring the 10, 15, 20-minute city and an evaluation of its use for sustainable urban design. Cities 131:103924. https://doi.org/10.1016/j.cities.2022.103924
35. McNeil N (2011) Bikeability and the 20-min neighborhood: how infrastructure and destinations influence bicycle accessibility. Transp Res Rec 2247(1):53–63
36. Moore-Cherry, N., Andersen, C. S., & Kayanan, C. M. (2023). Engaging 15-Minute Cities as a New Development Model: The Potential of Waterford City. In S. Flynn & R. Hayes (Eds.), Urban Planning for the City of the Future: A Multidisciplinary Approach (pp. 139–160). Emerald Publishing Limited. https://doi.org/10.1108/978-1-80455-215- 520231008
37. Moreno C, Allam Z, Chabaud D, Gall C, Pratlong F (2021) Introducing the "15-minute city": sustainability, resilience and place identity in future post-pandemic cities. Smart Cities 4(1):93–111. https://doi.org/10.3390/smartcities4010006
38. Murgante B, Valluzzi R, Annunziata A (2024) Developing a 15-minute city: evaluating urban quality using configurational analysis. The case study of Terni and Matera, Italy. Appl Geogr 162:103171
39. Murtagh EM, Murphy MH, Boone-Heinonen J (2010) Walking–the first steps in cardiovascular disease prevention. Curr Opin Cardiol 25(5):490
40. Nieuwenhuijsen MJ (2021) New urban models for more sustainable, liveable and healthier cities post covid19; reducing air pollution, noise and heat island effects and increasing green space and physical activity. Environ Int 157:106850
41. Pan Q, Jin Z, Liu X (2020) Measuring the effects of job competition and matching on employment accessibility. Transp Res Part D: Transp Environ 87:102535
42. Papas T, Basbas S, Campisi T (2023) Urban mobility evolution and the 15-minute city model: from holistic to bottom-up approach. Transp Res Procedia 69:544–551. https://doi.org/10.1016/j.trpro.2023.02.206
43. Pekdemir MI, Altintasi O, Ozen M (2024) Assessing the impact of public transportation, bicycle infrastructure, and land use parameters on a small-scale bike-sharing system: a case study of Izmir, Türkiye. Sustain Cities Soc 101:105085
44. Pozoukidou G, Chatziyiannaki Z (2021) 15-minute city: decomposing the new urban planning eutopia. Sustainability 13(2):928
45. Pritchard JP, Tomasiello DB, Giannotti M, Geurs K (2019) Potential impacts of bike-and-ride on job accessibility and spatial equity in São Paulo, Brazil. Transp Res Part A: Policy Pract 121:386–400
46. Rhoads D, Solé-Ribalta A, Borge-Holthoefer J (2023) The inclusive 15-minute city: walkability analysis with sidewalk networks. Comput Environ Urban Syst 100:101936. https://doi.org/10.1016/j.compenvurbsys.2022.101936
47. Sandoval R, Van Geffen C, Wilbur M, Hall B, Dubey A, Barbour W, Work DB (2021) Data driven methods for effective micromobility parking. Transp Res Interdiscip Perspect 10:100368
48. Staricco L (2022) 15-, 10- or 5-minute city? A focus on accessibility to services in Turin, Italy. J Urban Mobil 2:100030. https://doi.org/10.1016/j.urbmob.2022.100030
49. Stowe EW, Hughey SM, Hallum SH, Kaczynski AT (2019) Associations between walkability and youth obesity: differences by urbanicity. Child Obes 15(8):555–559
50. Wang F, Boros S (2021) The effect of daily walking exercise on sleep quality in healthy young adults. Sport Sci Health 17:393–401
51. Weng M, Ding N, Li J, Jin X, Xiao H, He Z, Su S (2019) The 15-minute walkable neighborhoods: measurement, social inequalities and implications for building healthy communities in urban China. J Transp Health 13:259–273. https://doi.org/10.1016/j.jth.2019.05.005

52. Zhang Y, Lin D, Mi Z (2019) Electric fence planning for dockless bike-sharing services. J Clean Prod 206(1):383–393
53. Zhang Y, Mi Z (2018) Environmental benefits of bike sharing: a big data-based analysis. Appl Energy 220(6):296–301
54. Zhang S, Zhen F, Kong Y, Lobsang T, Zou S (2023) Towards a 15-minute city: a network-based evaluation framework. Environ Plan B: Urban Anal City Sci 50(2):500–514. https://doi.org/10.1177/23998083221118570
55. Zhanga Y, Lin D, Liu XC (2019) Biking islands in cities: an analysis combining bike trajectory and percolation theory. J Transp Geogr 80(10):102497

Chapter 7
Plan Electric Geofences for Dockless Bike-Sharing

7.1 Introduction

As a form of sustainable transport, bike-sharing is becoming increasingly popular in cities around the world. This popularity can be attributed to the association of bike-sharing with various socioeconomic and environmental benefits, such as a reduction in CO_2 emissions, improvements in public health, and a decline in traffic congestion and noise pollution. These benefits arise from providing alternatives to car commuting and encouraging greater use of public transit [7, 23, 34]. We are now witnessing a new generation of bike-sharing, which is currently emerging in China and expanding globally. Prior to these dockless services, bikes have to be docked at stations. However, with the new service, bikes can be unlocked and paid for via a smartphone and then picked up and left anywhere at users' convenience. Dockless bike-sharing is developing at an unprecedented rate. As of July 2017, approximately 16 million dockless bikes were available in China, generating 50 million orders per day from 106 million users [40].

The rapid development of dockless bike-sharing significantly amplifies the various benefits of bike-sharing in supporting the achievement of sustainable cities, but it also generates new urban challenges. Among these issues, the most serious is arguably inappropriate parking behaviour. With the convenience of 'docklessness' or 'free-to-leave' policies, a notable number of users park bikes in places that are unsuitable as parking spaces (e.g., on pedestrian streets, directly adjacent to metro entrances, or inside gated communities), thereby causing negative impacts, such as violating pedestrian right-of-way, obstructing the movement of metro users, or making bikes less accessible to subsequent users.

An electric fence (or geo-fence) is traditionally defined as a barrier designed to protect economically and ecologically sensitive areas. Most electric fences are currently used for agricultural purposes and other forms of animal control [32]. Today, the Chinese government and major bike-sharing service providers regard electronic fence policies and technology as the primary solution for regulating inappropriate

Y. Zhang, *Understanding Shared Micromobility in China Using Massive User-Generated Trip Data*, Urban Sustainability,
https://doi.org/10.1007/978-981-96-8492-2_7

parking behaviour. These systems prevent users from parking bikes in prohibited areas and guide them to designated parking zones. An electric fence for dockless bike-sharing is a predetermined 'virtual fence' without a physical installation. The system operates using GPS or Bluetooth signals to determine the proximity between bikes and designated zones. Users who park bikes outside the allowed areas cannot lock them and will continue to incur charges. Additionally, riders are guided via their app or navigation prompts to the appropriate parking locations. The electric fence policy and technology have been recommended in several key governmental documents, such as *The National Guidance to Encourage and Regulate the Development of the Internet-based Dockless Bike-sharing Service*, issued by the Chinese Ministry of Transport in August 2017. This technology and policy have also been piloted in several cities in China (e.g., Beijing, Chengdu, and Hangzhou) since early 2017.

As a first effort to discuss electric fence technology and policy, this chapter primarily focuses on developing a series of methods to support electric fence planning for dockless bike-sharing services. It then applies these methods in a case study of Shanghai using a large dataset of bike trips provided by Mobike. The key method used to support electric fence planning is the location-allocation model, which has been widely applied in the planning of various urban facilities, including bike stations. Electric fence planning can be regarded as an optimisation problem, which can be addressed using the location allocation model. Based on a certain number of electric fences and the distribution of parking demand, the goal of the location-allocation model is to position the electric fences in the most efficient way to meet parking demand. The main data used in this chapter consists of user-generated bike trip data, which can be considered a form of urban big data, facilitating the accurate delineation of users' parking demand. Mobike, the largest dockless bike-sharing company in the world, and Shanghai, the largest bike-sharing market globally, are analysed in our case study, which enhances the representativeness of our research.

7.2 Literature Review

7.2.1 Dock-Based Bike-Sharing Analysis

The studies in the first domain include those that characterise bike-sharing through various analyses. For example, Wood et al. [43] visualised the dynamics of London's bike-sharing scheme using flow maps, while Zaltz Austwick et al. [44] employed visualisation, descriptive statistics, and spatial and network analysis tools to explore usage in five cities worldwide. Beecham et al. [3] proposed a new technique for classifying commuting behaviours, incorporating various spatial analysis algorithms and visual analytics techniques. Additionally, Caulfield et al. [7] examined the usage patterns of a bike-sharing programme in Cork, a medium-sized city in Ireland. This research provided insights into the dynamics of a relatively small bike-sharing scheme and assesses how bike-sharing has offered citizens a new transportation alternative. Researchers have also been particularly interested in understanding the factors that influence bike-sharing, such as the built environment, weather, and socioeconomic demographics [1, 9, 11]. Most previous studies have focused on the analysis of individual cities, such as London [26, 30, 43], Washington, D.C. [5, 6], Toronto and Montreal [14, 16], and Hangzhou and Zhongshan [8, 35, 47]. However, some studies have conducted comparative analyses of bike-sharing systems in different cities. These comparisons are based on numerous factors, such as the number of subscribers, stations, and bikes, modal share changes, connectivity, and flows [25, 44, 45].

The second domain of dock-based bike-sharing service analysis includes mathematical models that focus on rebalancing. Balancing bike-sharing stations to prevent them from becoming either too full or too empty is a critical issue. Several studies have addressed the rebalancing challenge. For example, using London's Barclays Cycle Hire programme as a case study, Pfrommer et al. [30] examined the efficient operation of shared mobility systems by combining intelligent routing decisions for staff-based vehicle redistribution with real-time price incentives for customers. Forma et al. [12] proposed a three-step mathematical programming-based heuristic method for the static repositioning problem. Erdoğan et al. [10] introduced the static bicycle rebalancing problem, aiming not only to determine the number of bikes to be allocated to each station but also to minimise the cost of stations served by only a few vehicles. Pal and Zhang [27] presented a novel mixed-integer linear programming model for solving real-life large-scale static rebalancing problems.

Facility location is one of the most important factors in strategic decision-making by the public sector. The studies in the third domain focus on the development of various methods to support bike-sharing facility location planning, particularly optimising the locations of bike stations. Supply-based bike facility planning has relied on quantitative models, such as the bicycle level of service [19] or the bicycle compatibility index [18], to assess roadway conditions for cyclists. Additional studies have sought to develop bicycle demand models to plan bike facilities, as the success of bike-sharing as a public transport mode largely depends on how well user demand is met [13, 15, 33].

Among the existing approaches to support bike station planning, the location-allocation model has proven to be highly effective in locating bike stations in relation to the distribution of potential demand [15]. It has been widely applied in the planning of various urban facilities [31]. In the context of public services, the most frequent applications include healthcare facilities [42], fire stations [21], schools [39], and waste disposals [20], among others. There are also studies that used the location-allocation model to optimise the locations of bike stations. For example, García-Palomares et al. [15] proposed a GIS-based method to calculate the spatial distribution of potential trip demand, locate stations using location-allocation models, determine station capacity, and define the characteristics of station demand. Frade and Ribeiro [13] adopted an optimisation method to design a bike-sharing system that maximises covered demand while considering the available budget as a constraint. Based on taxi trajectory data, Park and Sohn [29] proposed a framework for efficiently locating bicycle-sharing stations to replace short automobile trips.

Although specific considerations for electric fence planning are required due to the characteristic of 'docklessness', which creates a significant difference between dockless bike-sharing services and dock-based services, existing studies on the facility location planning for dock-based bike-sharing provide a valuable reference for the study in this chapter. From these studies, it is evident that certain approaches have the potential to be applied here, such as the bicycle level of service (BLOS) [19], the bicycle compatibility index [18], and the location-allocation model [15]. However, it should be noted that different methods require different datasets. Taking BLOS as an example, this measure is a function of per-lane motor vehicle traffic volume, vehicle speed, traffic mix, potential cross-street traffic generation, pavement surface condition, and pavement width for cycling [19]. These data are currently difficult to obtain in Shanghai. The main data we have obtained is user-generated bike trip data, which can be regarded as a form of urban big data, facilitating the accurate delineation of users' parking demand. Considering data availability, the location-allocation model, as a demand-based method, is chosen as our primary method to support electric fence planning for dockless bike-sharing services.

7.2.2 Dockless Bike-Sharing Service Analysis

The phenomenon of dockless bike-sharing services is relatively new, emerging in recent years, and relevant trip data are not easily obtainable due to various reasons, such as privacy concerns and business confidentiality. As a result, there are very few studies on dockless bike-sharing services. For example, Bao et al. [2] proposed a data-driven approach to develop bike lane construction plans based on bike trip data. Using big data techniques, Zhang and Mi [46] estimated the impacts of bike-sharing on energy use, and carbon dioxide and nitrogen oxide emissions in Shanghai from a spatiotemporal perspective. They found that, in 2016, bike-sharing in Shanghai saved 8,358 tonnes of petrol and reduced CO_2 and NO_X emissions by 25,240 and 64 tonnes, respectively. Tian et al. [41] employed a hybrid single-valued neutrosophic multi-criteria group decision-making (MCGDM) approach with quality function deployment (QFD) to support market segment evaluation and the selection of bike-sharing services. Another study by Tian et al. [41] presented a hybrid fuzzy multi-criteria decision-making approach to enhance knowledge on smart bike-sharing services. Shen et al. [36] analysed the usage patterns of dockless bike-sharing in Singapore using GPS data from one of the largest bike-sharing operators in the country. Pan et al. [28] focused on rebalancing the distribution of dockless bikes in cities by using an algorithm called loss-reduced reinforcement pricing (LRP). Liu et al. [22] addressed the problem of inferring dockless bike distribution in new cities by drawing insights from cities where dockless bikes are already present. To explore factors affecting the shift to dockless bikes, a binary logistic model was established for the modal shift from primarily substituted modes, namely walking and bus riding. Zhou and Ni [48] investigated factors affecting the dockless bike-sharing modal shift in metro commuting. They found that the time spent walking or riding a bus is the most critical factor. Additionally, good riding skills, non-motorised vehicle ownership, and metro locations outside the city centre have significantly positive effects on the modal shift. Based on a series of stakeholder interviews and media analysis, Spinney and Lin [38] concluded that dockless bike-sharing represents both a retrenchment and extension of existing exploitative capitalist relations. However, to the best of our knowledge, no study has yet discussed facility location planning for dockless bike-sharing services when the research was conducted.

7.3 Methodology

In this chapter, electric fences are regarded as a type of urban facility designed to accommodate the increasing parking demand for dockless bikes. Electric fence planning is conceptualised as an optimisation problem that can be addressed using the location-allocation model. Given a predetermined number of electric fences and the distribution of parking demand, the objective of the location-allocation model is to determine the most efficient placement of electric fences to best meet the demand. Our methodological framework consists of three main steps. First, we gather three key data sources: the road network, parking demand, and potential locations for electric fences. Second, we apply the location-allocation model to plan the placement of electric fences, using these data sources as inputs. Finally, we refine the selection of precise locations for electric fences using a method specifically proposed in this chapter. Notably, this final step, which is not part of the traditional location-allocation model, enhances the efficiency and accuracy of site selection for electric fences. The following sections provide a detailed explanation of each methodological step.

7.3.1 Construct Road Network

In this chapter, all bikeable roads in Shanghai are extracted from OpenStreetMap using the analytical package OSMnx. OpenStreetMap is a free, editable map of the world, collaboratively built by volunteers and released under an open-content licence. It has become one of the most widely used sources of free geographic data. OSMnx is a tool designed for downloading administrative boundary shapes and street networks from OpenStreetMap. By configuring parameters in OSMnx—specifically setting the road type to 'bikeable' and the administrative boundary to 'Shanghai'—we can efficiently extract bikeable roads in Shanghai for our research purposes. Once these bikeable roads are obtained, we construct the bike road network using ArcGIS.

7.3.2 Extract Parking Demand

The spatial distribution of parking demand is a fundamental element that must be determined for electric fence planning. In this study, each trip in the dataset includes its origin and destination (OD). Consequently, parking demand can be directly extracted from users' actual parking behaviour based on the bike trip dataset. Initially, we divide the study area into uniform urban cells, for example, each measuring 50 m by 50 m. We then count the number of ODs within each cell, using this as a representation of the cell's parking demand.

7.3.3 Select Location Candidates for Electric Fences

Selecting location candidates for electric fences is a prerequisite for the location-allocation model. High parking demand in an urban cell typically indicates that users are more likely to park bikes there, and this cell may also have sufficient space for bike parking. Therefore, parking demand is a critical factor to consider when selecting urban cells as location candidates for electric fences. An urban cell with several ODs exceeding a predetermined threshold is selected as a location candidate. It should be noted that other factors, such as planning constraints, the availability of relevant urban infrastructure, and the willingness of local residents, should also be considered in the selection process, assuming the relevant data are available.

7.3.4 Select Locations for Electric Fences

Based on a predetermined number of electric fences and the distribution of parking demand, the goal of the location-allocation model is to place the electric fences in the most efficient way to meet the parking demand. The location-allocation model has been widely applied in the planning of various urban facilities [31]. Several methods are commonly used to solve the location-allocation model, such as minimising impedance, maximising coverage, and minimising facilities. The choice of method depends on the objectives of stakeholders, including policymakers or service providers. The minimisation of impedance may be selected if the objective is to minimise the overall distance users need to travel to reach the designated electric fences; maximising coverage may be chosen if the aim is to allow more citizens to use the selected electric fences; and minimising facilities may be preferred if the construction budget is limited and the goal is to determine the minimum number of electric fences required to meet a certain amount of parking demand. In this study, we solve the location-allocation model using the maximise coverage method. To explain the location-allocation model more clearly, we first introduce the following notations:

i index of demand locations (i.e., urban cells), i=1,2,…, n
j index of location candidates for electric fences, j=1, 2, …, m
d_{ij} the network distance between i and j
D the impedance cut-off considered for the network distance
N_i a set of location candidate j within a distance D from demand location i
u_i the number of ODs in the i th demand location (amount of demand)
p the number of locations to be selected as electric fences (predetermined)
$x_i = 1$, if the demand location i is satisfied; 0, otherwise
$y_j = 1$, if the location candidate j is selected as an electric fence; 0, otherwise

The objective of the *maximise coverage* method to solve the location-allocation model is to maximise the demand covered by p electric fences. The mathematical model is formulated as follows:

$$\text{Maximise}: \sum_{i=1}^{n} u_i x_i \tag{7.1}$$

$$\sum_{j \in N_i} y_j \geq x_i \tag{7.2}$$

$$\sum_{j=1}^{m} y_j = p \tag{7.3}$$

$$x_i, y_j = (0, 1) \tag{7.4}$$

$$N_i = \{j | d_{ij} \leq D\} \tag{7.5}$$

To achieve its objective, the *maximise coverage* method seeks to maximise the demand coverage of electric fences. At least one electric fence must be within D from i to meet the i's parking demand.

7.3.5 Determine Capacities of Electric Fences

Spatial variations are observed in the capacities of electric fences, depending on the characteristics of the urban area they cover. For example, the capacity of an electric fence planned in a less-traffic area should be lower than that of one planned around a metro station, a tourist attraction, or a commercial street. Considering that the location-allocation model provides the amount of potential parking demand allocated to each electric fence, converting this demand into the number of bikes that can be docked at each fence is straightforward: the total number of bikes can be distributed among the different electric fences in proportion to the demand allocated to each fence [15]. For instance, if the total number of bikes is 1,200, the total parking demand is 10,000 ODs, and the demand allocated to an electric fence is 100 ODs, the capacity of this electric fence would be calculated as 1200 * (100/10,000) = 12 bikes. However, in practice, capacities are more likely to be designed according to fixed levels (e.g., 10, 20, or 50 bikes per electric fence) rather than non-fixed values (e.g., 7, 13, or 29 bikes per fence). Therefore, after determining the number of bikes based on the covered parking demand, we further adjust the capacities of the electric fences according to predefined capacity levels designed by relevant stakeholders. In the above example, the capacity of this electric fence might be determined as 10 bikes, rather than 12.

7.3.6 Recommend Accurate Places for Electric Fences

After selecting the locations for electric fences, the locations (i.e., uniform cells) where the electric fences should be placed are determined. However, the most accurate site within a selected location for setting up an electric fence must be identified, based on the size of the electric fence, which is much smaller than the size of an urban cell.

In this context, we propose a method for determining accurate sites that exhibit a high density of parking demand within a selected location. A high density of parking demand refers to an area with a large number of ODs in a relatively small space. This site is then recommended as the most suitable location for setting up an electric fence. The method is outlined as follows.

First, we cluster the ODs within the selected location (i.e., the chosen urban cell for setting up an electric fence) using DBSCAN (Density-Based Spatial Clustering of Applications with Noise). DBSCAN can group points that are densely packed (i.e., points with many nearby neighbours) into a cluster, while points that lie alone in low-density regions are marked as outliers (with their nearest neighbours too far away). DBSCAN requires only two parameters: *Eps*, the search radius (e.g., 3 m) of the neighbourhood, and *MinPts*, the minimum number of ODs in an Eps-neighbourhood (e.g., 2 ODs).

Next, we identify the cluster containing the largest number of ODs among all the clusters identified for each selected location.

We then calculate the mean centre of the cluster with the largest number of ODs. The mean centre represents the geographic centre (or the centre of concentration) for a set of ODs.

Finally, we delineate the minimum bounding rectangle of the cluster with the largest number of ODs. The minimum bounding rectangle is the smallest rectangle that encloses a set of ODs and is regarded as an approximation of the shape of an electric fence.

The mean centre and minimum bounding rectangle of the cluster with the largest number of ODs in a selected location can help us efficiently and accurately determine suitable sites for setting up an electric fence. The government or service provider can then proceed with the installation of the electric fences, taking into account other social, economic, and institutional considerations. Figure 7.1 illustrates an example of the electric fence planning process. In this example, the study area consists of a 5 × 5 grid of locations (i.e., uniform cells), and the objective is to plan one electric fence.

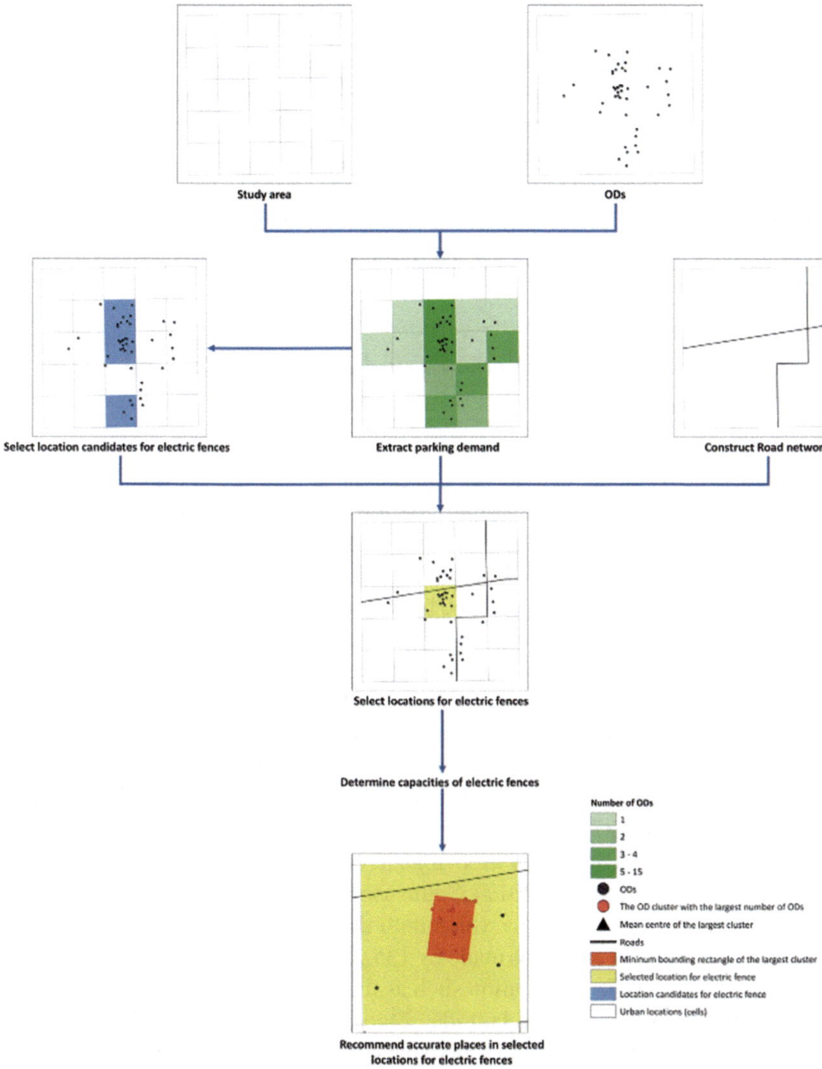

Fig. 7.1 Illustration of the electric fence planning process

7.4 Case Study

7.4.1 Study Area

With an area of 6,341 km^2 and a population of 24.3 million, Shanghai is one of the largest cities in the world and China's economic centre (Fig. 7.2a). Shanghai's GDP in 2016 was 2.8 trillion Yuan, accounting for 3.6% of China's total GDP. The Huangpu, Putuo, Hongkou, Jingan, Changning, Yangpu, and Xuhui districts are considered the traditional central area of Shanghai's sixteen districts. Lujiazui, Shanghai's CBD (central business district), is located in the Pudong New District, adjacent to the traditional central area. As of July 2017, Shanghai had 1.5 million dockless bikes, making it the largest bike-sharing market in the world.

7.4.2 Constructed Road Network and Extracted Parking Demand

We construct the bikeable road network in Shanghai (Fig. 7.2b) using OSMnx and ArcGIS. The total length of the roads is 26,900 km. This road network comprises 116.9 thousand street segments and 80.9 thousand street nodes. Figure 7.3 clearly illustrates the high density of roads in the central area.

The data used in this study is provided by Mobike, a dockless bike-sharing service provider that aims "to fulfil urban short trips—anytime, to any authorised parking destination—by combining innovation and today's IoT (Internet of Things) technology" [24]. This dataset is formally authorised for use in research purposes. As of March 2017, Mobike had more than 4 million red-framed bikes in nearly 80 cities worldwide. The company receives approximately 20 million orders per day, accounting for 56.56% of the total market, making it the largest dockless bike-sharing company in the world [37].

We obtain a bike trip dataset of users registered in Shanghai, covering a period of fifteen days (15–30 September 2017). Some users also generate trip records outside Shanghai. After selecting trips for which both the origin and destination are within Shanghai, we obtain 777,896 trips made by 135,239 users using 298,998 bikes. Each trip record contains basic information such as the order identification (ID), user ID, bike ID, start time, longitude and latitude of the origin, end time, and longitude and latitude of the destination. Therefore, bike parking demand can be directly extracted from the users' ODs.

We divide Shanghai into uniform cells, each with a width of 50 m and a length of 50 m, and then count the number of ODs in each cell. Among the 2,718,296 urban cells in Shanghai, 186,881 cells have been used as a parking space at least once. The average number of ODs per cell is 8.3, with a standard deviation of 18.1. As shown in Fig. 7.2c, most of the parking demand is concentrated within the main ring

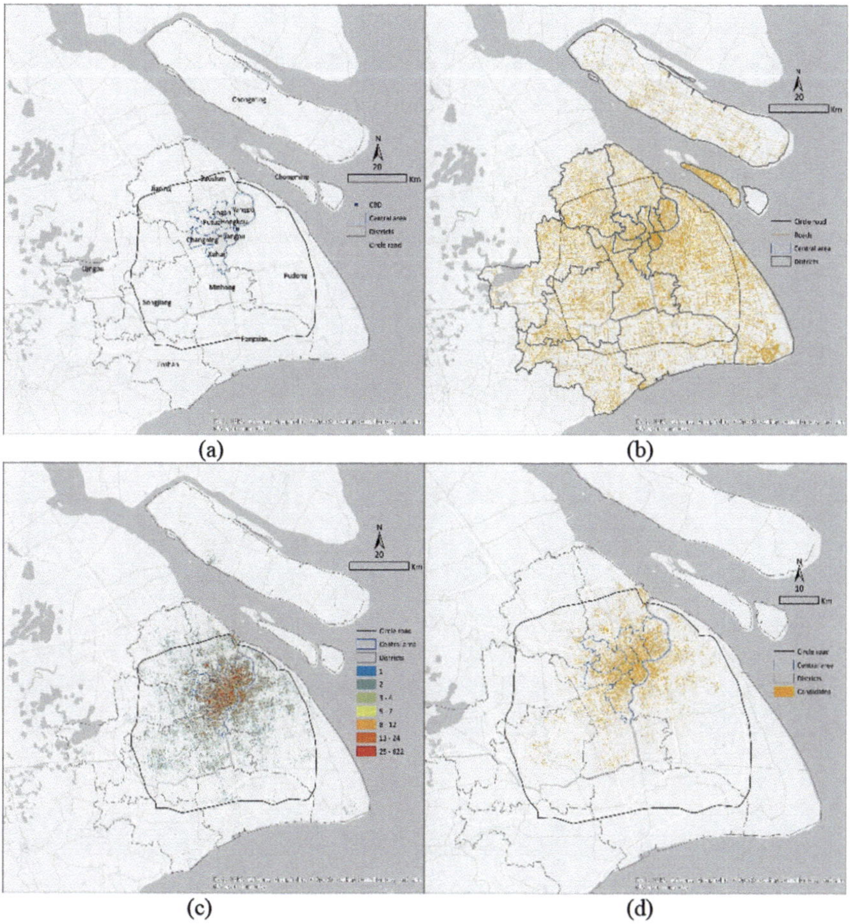

Fig. 7.2 Study area, road network, parking demand, and location candidates. Note: Map of Shanghai (**a**), bikeable road network in Shanghai (**b**), bike parking demand of users (**c**), and distribution of location candidates for electric fences (**d**)

road. Particularly high parking demand is observed in the central areas, such as the Huangpu, Jiangan, and Yangpu districts.

7.4.3 Selected Location Candidates

Due to the lack of other supporting data, this chapter selects location candidates for electric fences based on the parking demand retrieved from our bike trip dataset. Specifically, we select all urban cells with a parking demand greater than 5 ODs as location candidates for electric fences. There are a total of 58,941 urban cells. The

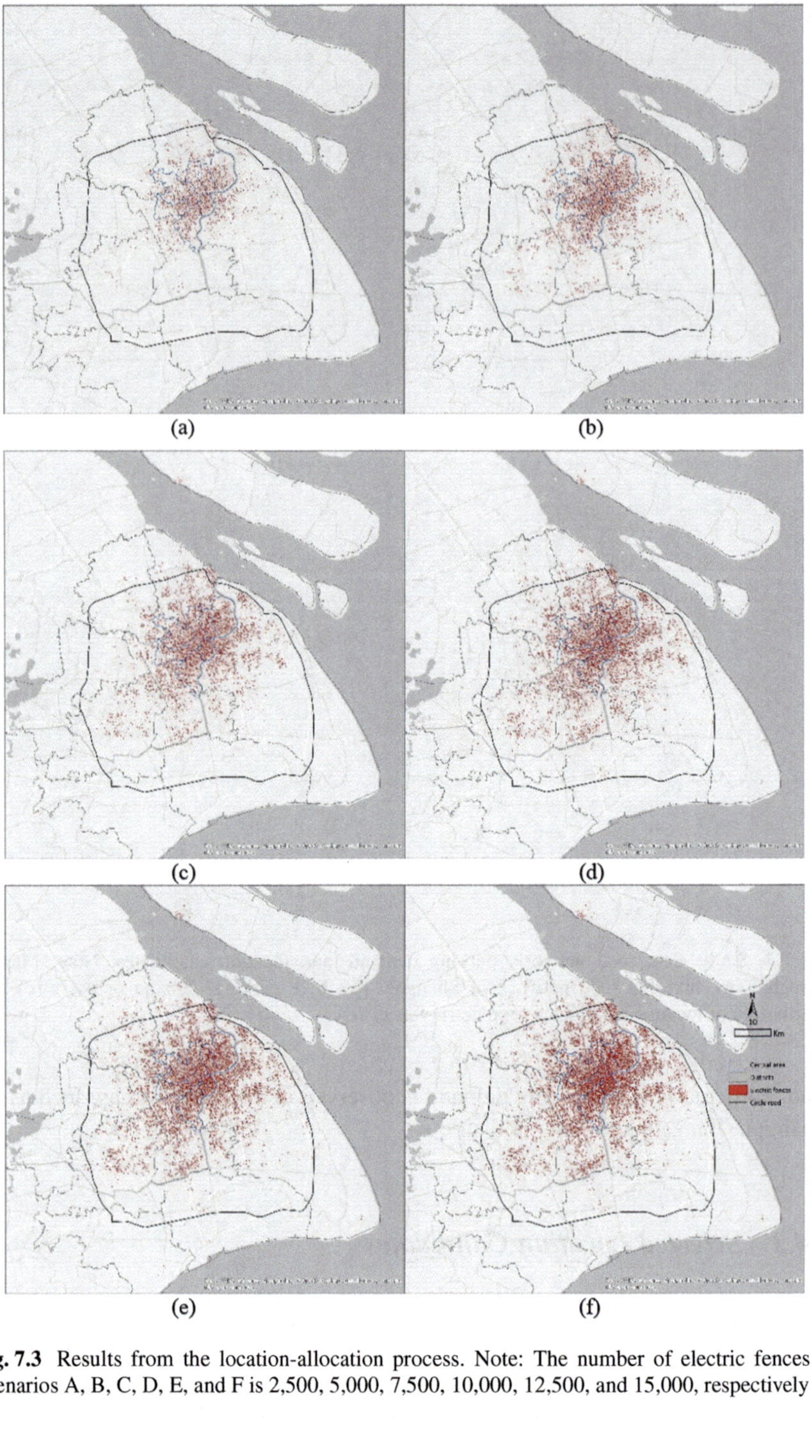

Fig. 7.3 Results from the location-allocation process. Note: The number of electric fences in Scenarios A, B, C, D, E, and F is 2,500, 5,000, 7,500, 10,000, 12,500, and 15,000, respectively

average number of ODs in each cell is 22.1, with a standard deviation of 27.5. As shown in Fig. 7.2d, most of the location candidates are within the main ring road and are especially concentrated in the central area. Excluding urban cells with a parking demand of fewer than 5 ODs allows the location-allocation model to run more efficiently. The number of selected location candidates (i.e., 58,941) is much larger than the expected number of electric fences, thus providing sufficient flexibility to identify an optimal solution for the location-allocation model.

7.4.4 Selected Locations for Electric Fences

We solve the location-allocation model using the *maximise coverage* method. In this method, the impedance cut-off considered for the network distance is 200 m, which is deemed to be a suitable distance for pedestrian access to bikes [15]. By adopting this cut-off, we assume that an electric fence can cover the demand within a network distance of 200 m.

The number of planned electric fences is determined based on the following considerations. First, we determine the smallest and largest number of planned electric fences by referring to two electric fence projects currently being tested in China. One project in Beijing has 20 electric fences with a capacity of 2,000 bikes (i.e., an average of 100 bikes per electric fence) [4], while another project in Hangzhou has 38 electric fences with a capacity of 1,000 bikes (i.e., an average of 26.3 bikes per electric fence) [17]. Referring to these capacities, we would require 2,990 and 11,369 electric fences to ensure that all 298,998 bikes could be docked at one of the planned electric fences. Second, existing studies typically present their results for the location-allocation model using several fixed values. For example, García-Palomares et al. [15] illustrated their results using three fixed numbers of bike stations: 100, 300, and 500 stations, Park and Sohn [29] presented their results using three fixed numbers of bike stations: 40, 80, and 120 stations. Therefore, in this chapter, we consider six scenarios to illustrate our results, based on the total number of planned electric fences: 2,500; 5,000; 7,500; 10,000; 12,500; and 15,000.

We use parking demand, location candidates, the road network, and the number of planned electric fences as inputs, and solve the location-allocation model using the network analysis function provided in ArcGIS. Figure 7.3 shows the results for the planned electric fences across the six scenarios. When the number of planned electric fences is 2,500, most are located in the central area and part of Pudong New District, which is adjacent to the central area. As the number of planned electric fences increases, the electric fences expand rapidly from the central area to the suburban areas, although most remain within the main ring road.

Table 7.1 shows diminishing returns with an increase in the number of planned electric fences.

When the number of electric fences reaches 2,500, 68.2% of parking demand is covered. An increase of 2,500 electric fences, from Scenario A to B, results in a significant rise in covered demand (from 68.2 to 85.0%). However, subsequent

Table 7.1 Percentages of covered parking demand

Scenarios	Number of planned electric fences					
	2,500	5,000	7,500	10,000	12,500	15,000
Percentages of covered demand (%)	68.2	85.0	91.8	95.0	96.3	96.4

increases in covered demand become less pronounced with further additions of electric fences. For example, an increase from Scenario B to C leads to a 6.8% rise in covered demand, while the addition of electric fences from 7500 to 10,000 only increases the covered demand by 3.2%. Scenario C (with 7,500 electric fences) is the first to cover more than 90% of parking demand, making it the optimal electric fence plan. This scenario ensures a high level of demand coverage with a relatively manageable number of electric fences. The following analysis is further illustrated based on this scenario.

We focus on the central area of Shanghai (Huangpu District) (Fig. 7.4) to better observe the spatial distribution of the planned electric fences. Each red urban cell represents a selected location for an electric fence. One endpoint of each line is connected to an urban cell, the parking demand of which will be covered by an electric fence at the other endpoint of the line. The colour of each line indicates the number of ODs covered by the electric fence it is linked to. Most of the parking demand is covered by a nearby electric fence. The majority of electric fences are located near roads, particularly at road junctions.

7.4.5 Determined Capacities of Electric Fences

In Scenario C, 7,500 electric fences are planned, with an average capacity of 40 bikes, ensuring that all 298,998 bikes can be docked at an electric fence. To determine the capacities of electric fences, the number of bikes to be docked at each fence is first allocated in proportion to the demand assigned to it. We then consider the following fixed capacity levels in this study: 10, 20, 50, 100, 150, and 200 or more bikes. For example, if the number of bikes to be docked at an electric fence based on the covered parking demand is 13, the capacity is set at 10 bikes. Table 7.2 shows that most electric fences have a capacity of 20 bikes, followed by those with capacities of 50 and 10 bikes. At least 286,480 bikes (95.8%) can be docked at one of the planned electric fences.

7.4.6 Recommended Accurate Places in Selected Locations

Appropriate values for two parameters in DBSCAN must be chosen when identifying the OD clusters at selected locations: *Eps*, the neighbourhood search radius, and

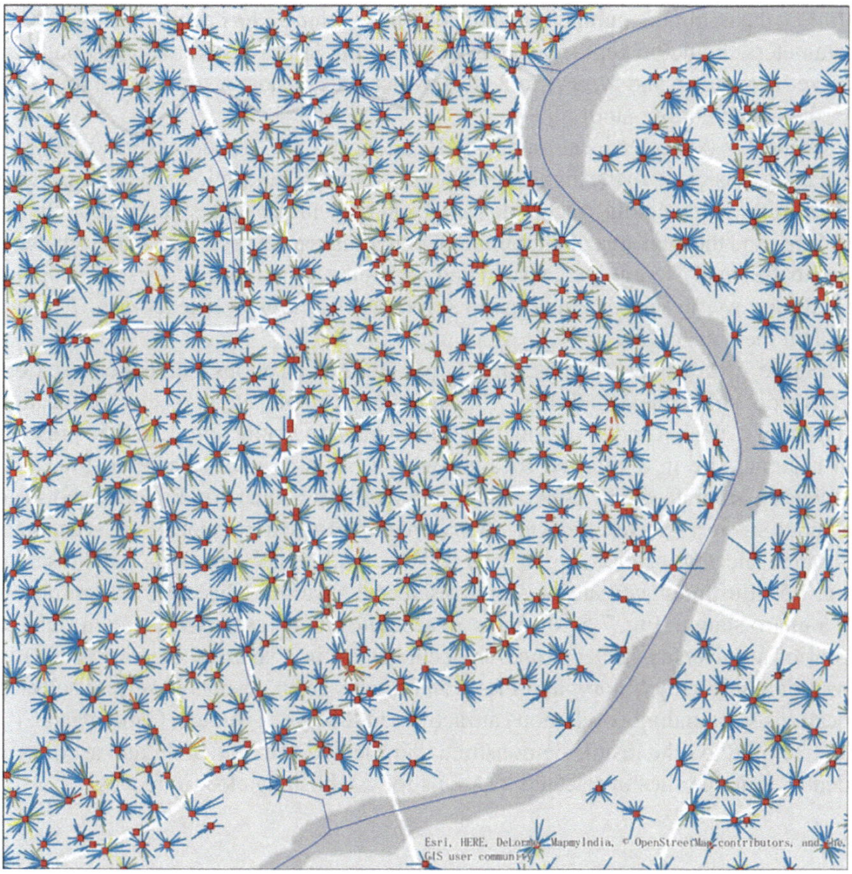

Fig. 7.4 Results in the Shanghai central area from Scenario C (7,500 electric fences)

Table 7.2 Capacities of the planned electric fences in Scenario C

	Number of bikes can be docked according to the covered demand						In total
	0–15	15–35	35–75	75–125	125–175	175–	
Fixed capacity levels	10	20	50	100	150	≥ 200	
Number of electric fences	1,956	2,576	2,024	654	184	106	7,500
Number of bikes can be docked	19,560	515,20	101,200	65,400	27,600	≥ 21,200	≥ 286,480

MinPts, the minimum number of demand points within an *Eps* neighbourhood. Both parameters should be kept small to avoid detecting large clusters with OD outliers. We set *Eps* to 3 m and *MinPts* to 2 ODs, which we consider a suitable combination for the context of our study.

Of the 7,500 selected locations, 6,770 locations (90.3%) are successfully identified as having clusters with the highest number of ODs in each location. The average area of the minimum bounding rectangle for the cluster with the largest number of ODs is 25.5 m^2, and the total area of all minimum bounding rectangles is 172.4 thousand m^2. Due to the small average area, visualising the results for the entire city is challenging. Therefore, we present our findings in several typical urban contexts (Fig. 7.5) and explain them as follows.

Figure 7.5a displays accurate sites in two selected locations near a metro station, a bus stop, and a main road; Fig. 7.5b shows an accurate site in a selected location in front of a cinema and a supermarket, near a main road junction; Fig. 7.5c presents an accurate site in a selected location near a bus stop and a hospital; Fig. 7.5d depicts an accurate site in a selected location within a green park; Fig. 7.5e shows an accurate site in a selected small location (left) adjacent to a city plaza, and an accurate site in a selected location (right) near a metro station and the China Art Museum; Fig. 7.5f illustrates accurate sites in two selected locations adjacent to the entrances of a metro station; Fig. 7.5g displays an accurate site in a selected location (left) in front of a university canteen, and an accurate site in a selected location (right) in front of a university library; and Fig. 7.5h presents accurate sites in several selected locations surrounding two famous attractions in Shanghai: Yuyuan Garden and City God Temple. As the results demonstrate, our method can efficiently and accurately identify suitable sites in selected locations for establishing electric fences.

7.5 Conclusions and Discussion

A new generation of bike-sharing services is emerging in China. Through these services, bikes can be unlocked and paid for using a smartphone, and then picked up and left anywhere at the user's convenience. The unprecedented rise of dockless bike-sharing services has brought significant socioeconomic and environmental benefits, but it has also introduced new urban challenges. One of the most pressing issues is inappropriate parking behaviour. To address this problem, the electric fence policy and technology are introduced to guide users towards parking bikes in designated areas. In this chapter, we propose a series of methods to support the planning of electric fences for dockless bike-sharing services. These methods are then applied in a case study of Shanghai, using a large dataset of bike trips. We employ the *maximise coverage* method to solve the location-allocation model and present the results across six scenarios. Finally, we use several methods (i.e., DBSCAN, minimum bounding rectangle, and the mean centre) to efficiently identify accurate sites that can be recommended for the installation of electric fences. It should be noted that the final electric fence plan should be developed in conjunction with other social (e.g., local citizens'

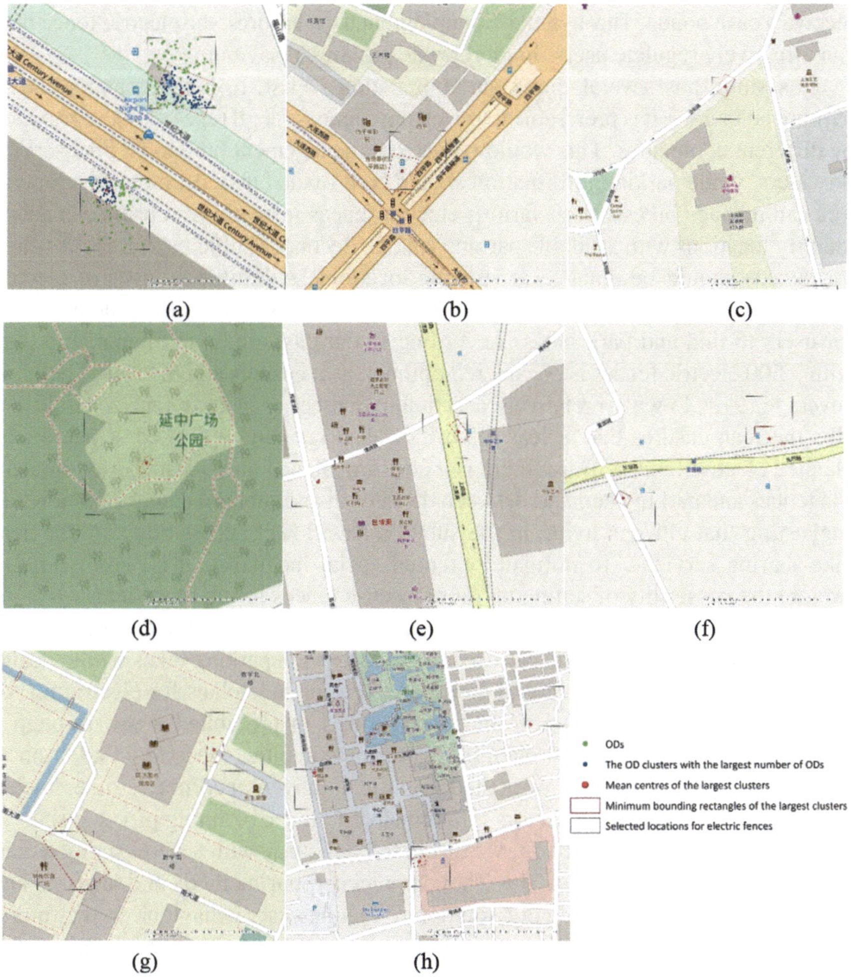

Fig. 7.5 Examples of recommended accurate sites in selected locations

willingness), economic (e.g., the cost of electric fences and their maintenance), and institutional (e.g., planning regulations) considerations.

In addition to the location plans for electric fences, technical and policy support is also required for the successful implementation of the electric fence plan. The electric fence system operates using GPS or Bluetooth signals to determine the proximity between bikes and designated zones. Bike riders are guided to appropriate parking locations via their app or navigation prompts. Moreover, service providers can introduce various incentives or punitive policies. For example, users who park bikes outside the electric fences will not be able to lock them and will continue to be charged, whereas users who park bikes correctly inside the electric fences may receive a cash bonus. Through these supplementary measures, the electric fence plan can effectively regulate users' inappropriate parking behaviour.

This study has several important implications. First, to better manage users' parking behaviour, it is preferable to use electric fences for all types of bikes provided by different companies. This would reduce the management burden on bike service providers, while parking information should be provided in real-time so that operators can manage bikes across various electric fences instantly, and users can easily identify locations with available parking spaces. Second, a sufficient number of electric fences should be planned at suitable locations. A limited number of electric fences would undermine the advantage of 'docklessness' and make it inconvenient for users to find and park bikes. According to our case study, in Shanghai, a plan with 7,500 electric fences is considered optimal, as it ensures a high level of demand coverage (91.8%) with a relatively acceptable number of electric fences. Additionally, the plan ensures that at least 95.8% of all bikes can be docked at one of the planned electric fences. Third, there is a clear imbalance in the distribution of electric fences and parking demand between the central and suburban areas of Shanghai, suggesting that citizens living in the suburbs have lower accessibility to dockless bike-sharing services. To mitigate potential social inequality, it is worthwhile to explore the possibility of setting up more electric fences in suburban areas.

This study can be improved and extended in several ways. First, we have only discussed plans for electric fences used to guide users in parking bikes within designated zones. Our discussion could be expanded to include plans for electric fences implemented to prevent users from parking bikes in prohibited areas, through a combination of various urban planning constraints in cities. Second, in addition to considering parking demand, we could also quantitatively incorporate other social, economic, and institutional factors into the electric fence planning process, such as the cost of electric fences. Third, we discussed our study with several senior scientists from Mobike in March 2018. They appreciated our methods and analysis and expressed interest in adopting our solution to regulate user behaviour. In the future, we may explore the possibility of applying our methods to support electric fence planning through collaboration with Mobike's research team. Finally, based on our methods, we could support electric fence planning for other service providers (e.g., ofo), assuming bike trip data were available. By using trip data provided by several major service providers, we could develop a comprehensive electric fence plan, ensuring that the established electric fences are accessible and free for all dockless bike-sharing services.

References

1. Bachand-Marleau J, Lee B, El-Geneidy A (2012) Better understanding of factors influencing likelihood of using shared bicycle systems and frequency of use. Transp Res Rec J Transp Res Board 2314:66–71. https://doi.org/10.3141/2314-09

2. Bao J, He T, Ruan S et al (2017) Planning bike lanes based on sharing-bikes' trajectories. In: Proceedings of the 23rd ACM SIGKDD international conference on knowledge discovery and data mining—KDD '17. ACM Press, New York, New York, USA, pp 1377–1386
3. Beecham R, Wood J, Bowerman A (2014) Studying commuting behaviours using collaborative visual analytics. Comput Environ Urban Syst 47:5–15. https://doi.org/10.1016/j.compenvurbsys.2013.10.007
4. Beijing Evening News (2017) Beijing is setting up electric fences in Chaoyang District. Beijing Evening News
5. Buck D, Buehler R, Happ P et al (2013) Are bikeshare users different from regular cyclists? Transp Res Rec J Transp Res Board 2387:112–119. https://doi.org/10.3141/2387-13
6. Buehler R (2012) Determinants of bicycle commuting in the Washington, DC region: the role of bicycle parking, cyclist showers, and free car parking at work. Transp Res Part Transp Environ 17:525–531. https://doi.org/10.1016/j.trd.2012.06.003
7. Caulfield B, O'Mahony M, Brazil W, Weldon P (2017) Examining usage patterns of a bike-sharing scheme in a medium sized city. Transp Res Part Policy Pract 100:152–161. https://doi.org/10.1016/j.tra.2017.04.023
8. Chen M, Wang D, Sun Y et al (2017) Service evaluation of public bicycle scheme from a user perspective. Transp Res Rec J Transp Res Board 2634:28–34. https://doi.org/10.3141/2634-04
9. El-Assi W, Salah Mahmoud M, Nurul Habib K (2017) Effects of built environment and weather on bike sharing demand: a station level analysis of commercial bike sharing in Toronto. Transportation 44:589–613. https://doi.org/10.1007/s11116-015-9669-z
10. Erdoğan G, Battarra M, Wolfler Calvo R (2015) An exact algorithm for the static rebalancing problem arising in bicycle sharing systems. Eur J Oper Res 245:667–679. https://doi.org/10.1016/j.ejor.2015.03.043
11. Faghih-Imani A, Eluru N, El-Geneidy AM et al (2014) How land-use and urban form impact bicycle flows: evidence from the bicycle-sharing system (BIXI) in Montreal. J Transp Geogr 41:306–314. https://doi.org/10.1016/j.jtrangeo.2014.01.013
12. Forma IA, Raviv T, Tzur M (2015) A 3-step math heuristic for the static repositioning problem in bike-sharing systems. Transp Res Part B Methodol 71:230–247. https://doi.org/10.1016/j.trb.2014.10.003
13. Frade I, Ribeiro A (2015) Bike-sharing stations: a maximal covering location approach. Transp Res Part Policy Pract 82:216–227. https://doi.org/10.1016/J.TRA.2015.09.014
14. Fuller D, Gauvin L, Kestens Y et al (2011) Use of a new public bicycle share program in Montreal, Canada. Am J Prev Med 41:80–83. https://doi.org/10.1016/j.amepre.2011.03.002
15. García-Palomares JC, Gutiérrez J, Latorre M (2012) Optimizing the location of stations in bike-sharing programs: a GIS approach. Appl Geogr 35:235–246. https://doi.org/10.1016/J.APGEOG.2012.07.002
16. Habib KN, Mann J, Mahmoud M, Weiss A (2014) Synopsis of bicycle demand in the City of Toronto: investigating the effects of perception, consciousness and comfortability on the purpose of biking and bike ownership. Transp Res Part Policy Pract 70:67–80. https://doi.org/10.1016/j.tra.2014.09.012
17. Hangzhou Daily (2017) The use of electric fences in Hanghzou. Hangzhou Dly
18. Harkey D, Reinfurt D, Knuiman M (1998) Development of the bicycle compatibility index. Transp Res Rec J Transp Res Board 1636:13–20. https://doi.org/10.3141/1636-03
19. Landis B, Vattikuti V, Brannick M (1997) Real-time human perceptions: toward a bicycle level of service. Transp Res Rec J Transp Res Board 1578:119–126. https://doi.org/10.3141/1578-15
20. List G, Mirchandani P (1991) An integrated network/planar multiobjective model for routing and siting for hazardous materials and wastes. Transp Sci 25:146–156. https://doi.org/10.1287/trsc.25.2.146
21. Liu N, Huang B, Chandramouli M (2006) Optimal siting of fire stations using GIS and ANT algorithm. J Comput Civ Eng 20:361–369. https://doi.org/10.1061/(ASCE)0887-3801(2006)20:5(361)
22. Liu Z, Shen Y, Zhu Y (2018) Inferring dockless shared bike distribution in new cities. In: Proceedings of the eleventh ACM international conference on web search and data mining—WSDM '18. ACM Press, New York, New York, USA, pp 378–386

23. Martens K (2007) Promoting bike-and-ride: the Dutch experience. Transp Res Part Policy Pract 41:326–338. https://doi.org/10.1016/J.TRA.2006.09.010
24. Mobike (2017) About - Mobike. https://mobike.com/uk/about. Accessed 30 Nov 2017
25. O'Brien O, Cheshire J, Batty M (2014) Mining bicycle sharing data for generating insights into sustainable transport systems. J Transp Geogr 34:262–273. https://doi.org/10.1016/j.jtrangeo.2013.06.007
26. Ogilvie F, Goodman A (2012) Inequalities in usage of a public bicycle sharing scheme: socio-demographic predictors of uptake and usage of the London (UK) cycle hire scheme. Prev Med 55:40–45. https://doi.org/10.1016/j.ypmed.2012.05.002
27. Pal A, Zhang Y (2017) Free-floating bike sharing: solving real-life large-scale static rebalancing problems. Transp Res Part C Emerg Technol 80:92–116. https://doi.org/10.1016/j.trc.2017.03.016
28. Pan L, Cai Q, Fang Z et al (2018) Rebalancing dockless bike sharing systems
29. Park C, Sohn SY (2017) An optimization approach for the placement of bicycle-sharing stations to reduce short car trips: an application to the city of Seoul. Transp Res Part Policy Pract 105:154–166. https://doi.org/10.1016/J.TRA.2017.08.019
30. Pfrommer J, Warrington J, Schildbach G, Morari M (2014) Dynamic vehicle redistribution and online price incentives in shared mobility systems. IEEE Trans Intell Transp Syst 15:1567–1578. https://doi.org/10.1109/TITS.2014.2303986
31. Rahman S, Smith DK (2000) Use of location-allocation models in health service development planning in developing nations. Eur J Oper Res 123:437–452. https://doi.org/10.1016/S0377-2217(99)00289-1
32. Reidy MM, Campbell TA, Hewitt DG (2008) Evaluation of electric fencing to inhibit feral pig movements. J Wildl Manag 72:1012–1018. https://doi.org/10.2193/2007-158
33. Rybarczyk G, Wu C (2010) Bicycle facility planning using GIS and multi-criteria decision analysis. Appl Geogr 30:282–293. https://doi.org/10.1016/J.APGEOG.2009.08.005
34. Shaheen S, Cohen A, Martin E (2013) Public bikesharing in North America. Transp Res Rec J Transp Res Board 2387:83–92. https://doi.org/10.3141/2387-10
35. Shaheen S, Zhang H, Martin E, Guzman S (2011) China's Hangzhou public bicycle. Transp Res Rec J Transp Res Board 2247:33–41. https://doi.org/10.3141/2247-05
36. Shen Y, Zhang X, Zhao J (2018) Understanding the usage of dockless bike sharing in Singapore. Int J Sustain Transp. https://doi.org/10.1080/15568318.2018.1429696
37. Sohu Financial (2017) The market share of Mobike is approaching 60%. http://business.sohu.com/20170517/n493354558.shtml. Accessed 30 Nov 2017
38. Spinney J, Lin W-I (2018) Are you being shared? Mobility, data and social relations in Shanghai's public bike sharing 2.0 sector. Appl Mobilities 3:66–83. https://doi.org/10.1080/23800127.2018.1437656
39. Teixeira J, Antunes A, Peeters D (2007) An optimization-based study on the redeployment of a secondary school network. Environ Plan B Plan Des 34:296–315. https://doi.org/10.1068/b31173
40. The Centre for Sharing Economy (2017) The research report on bike-sharing employment
41. Tian Z-P, Wang J-Q, Zhang H-Y (2018) Hybrid single-valued neutrosophic MCGDM with QFD for market segment evaluation and selection. J Intell Fuzzy Syst 34:177–187. https://doi.org/10.3233/JIFS-171055
42. Verter V, Lapierre SD (2002) Location of preventive health care facilities. Ann Oper Res 110:123–132. https://doi.org/10.1023/A:1020767501233
43. Wood J, Slingsby A, Dykes J (2011) Visualizing the dynamics of London's bicycle hire scheme. Cartogr Int J Geogr Inf Geovisualization 46:239–251. https://doi.org/10.3138/carto.46.4.239
44. Zaltz Austwick M, O'Brien O, Strano E, Viana M (2013) The structure of spatial networks and communities in bicycle sharing systems. PLoS ONE 8:e74685. https://doi.org/10.1371/journal.pone.0074685
45. Zhang L, Zhang J, Duan ZY, Bryde D (2015) Sustainable bike-sharing systems: characteristics and commonalities across cases in urban China. J Clean Prod 97:124–133. https://doi.org/10.1016/j.jclepro.2014.04.006

46. Zhang Y, Mi Z (2018) Environmental benefits of bike sharing: a big data-based analysis. Appl Energy 220:296–301. https://doi.org/10.1016/j.apenergy.2018.03.101
47. Zhang Y, Thomas T, Brussel M, van Maarseveen M (2017) Exploring the impact of built environment factors on the use of public bikes at bike stations: case study in Zhongshan, China. J Transp Geogr 58:59–70. https://doi.org/10.1016/j.jtrangeo.2016.11.014
48. Zhou S, Ni Y (2018) Effects of dockless bike on modal shift in metro commuting: a pilot study in Shanghai

GPSR Compliance

The European Union's (EU) General Product Safety Regulation (GPSR) is a set of rules that requires consumer products to be safe and our obligations to ensure this.

If you have any concerns about our products, you can contact us on ProductSafety@springernature.com

In case Publisher is established outside the EU, the EU authorized representative is:

Springer Nature Customer Service Center GmbH
Europaplatz 3
69115 Heidelberg, Germany

Batch number: 10725825

Printed by Printforce, the Netherlands